NOVEL AI AND DATA SCIENCE ADVANCEMENTS FOR SUSTAINABILITY IN THE ERA OF COVID-19

NOVEL AI AND DATA SCIENCE ADVANCEMENTS FOR SUSTAINABILITY IN THE ERA OF COVID-19

Edited by

VICTOR CHANG
Full Professor, Department of Operations and Information Management, Aston Business School, Aston University, Birmingham, United Kingdom

MOHAMED ABDEL-BASSET
Associate Professor and Head of Department, Zagazig University, Egypt

MUTHU RAMACHANDRAN
CEO & Educational and Research Consultant, Integrated Cloud Solutions, Leeds, United Kingdom

NICOLAS G. GREEN
Bioelectronics and Microfluidity, Electronics and Computer Science, University of Southampton, United Kingdom

GARY WILLS
Cyber Physical Research Group, Electronics and Computer Science, University of Southampton, United Kingdom

ELSEVIER

ACADEMIC PRESS
An imprint of Elsevier

Academic Press is an imprint of Elsevier
125 London Wall, London EC2Y 5AS, United Kingdom
525 B Street, Suite 1650, San Diego, CA 92101, United States
50 Hampshire Street, 5th Floor, Cambridge, MA 02139, United States
The Boulevard, Langford Lane, Kidlington, Oxford OX5 1GB, United Kingdom

ISBN 978-0-323-90054-6

For information on all Academic Press publications
visit our website at https://www.elsevier.com/books-and-journals

Publisher: Stacy Masucci
Acquisitions Editor: Rafael E. Teixeira
Editorial Project Manager: Tracy Tufaga
Production Project Manager: Swapna Srinivasan
Cover Designer: Victoria Pearson

Typeset by STRAIVE, India

Contents

Contributors

Mohamed Aleem Ali
Cybersecurity, Information Systems and AI Research Group, School of Computing, Engineering and Digital Technologies, Teesside University, Middlesbrough, United Kingdom

T. Ananth Kumar
IFET College of Engineering, Gangarampalaiyam, India

S. Arunmozhiselvi
St. John the Baptist University, Mangochi, Malawi

Alka Bali
University Institute of Pharmaceutical Sciences, UGC Center of Advanced Study, Panjab University, Chandigarh, India

Nishu Bali
Department of Computer Sciences, Chitkara University, Patiala, India

Rachid Benlamri
Department of Software Engineering, Faculty of Engineering, Lakehead University, Thunder Bay, ON, Canada

Prasad Calyam
Electrical Engineering and Computer Science Department, University of Missouri-Columbia, Columbia, MO, United States

Victor Chang
Department of Operations and Information Management, Aston Business School, Aston University, Birmingham, United Kingdom

Chinmay Kumar Das
Kalinga Institute of Industrial Technology, Deemed to be University, Bhubaneswar, India

Rahul Kumar Dubey
Robert Bosch Engineering and Business Solutions Private Limited, Bengaluru, Karnataka, India

Ashish Garg
Walmart Global Tech, Bentonville, AR, United States

Amin Gasmi
Société Francophone de Nutrithérapie et de Nutrigénétique Appliquée; Interuniversity Laboratory of Motor Biology, Claude Bernard University, Villeurbanne, France

Ragib Hasan
Department of Computer Science, University of Alabama at Birmingham, Birmingham, AL, United States

Raiful Hasan
Department of Computer Science, University of Alabama at Birmingham, Birmingham, AL, United States

Alamgir Hossain
Academic Affairs & Research, Cambodia University of Technology and Science (CamTECH), Phnom Penh, Cambodia

Mauro Lemus Alarcon
Electrical Engineering and Computer Science Department, University of Missouri-Columbia, Columbia, MO, United States

Prasant Kumar Nanda
Kalinga Institute of Industrial Technology, Deemed to be University, Bhubaneswar, India

Chandra Bhanu Nayak
Kalinga Institute of Industrial Technology, Deemed to be University, Bhubaneswar, India

Roland Oruche
Electrical Engineering and Computer Science Department, University of Missouri-Columbia, Columbia, MO, United States

Ashish Pandey
Electrical Engineering and Computer Science Department, University of Missouri-Columbia, Columbia, MO, United States

N. Raghavendra Rao
FINAIT Consultancy Services, Chennai, India

G. Rajakumar
Francis Xavier Engineering College, Tirunelveli, India

Rojalin Sahu
Kalinga Institute of Industrial Technology, Deemed to be University, Bhubaneswar, India

R. Sharon Jebaleela
Francis Xavier Engineering College, Tirunelveli, India

R. Shree Charran
Department of Management Studies, Indian Institute of Science, Bengaluru, Karnataka, India

Sukanta Chandra Swain
Kalinga Institute of Industrial Technology, Deemed to be University, Bhubaneswar, India

Snigdha Tripathy
Kalinga Institute of Industrial Technology, Deemed to be University, Bhubaneswar, India

CHAPTER 1

Deep learning-based hybrid models for prediction of COVID-19 using chest X-ray

R. Shree Charran[a] and Rahul Kumar Dubey[b]
[a]Department of Management Studies, Indian Institute of Science, Bengaluru, Karnataka, India
[b]Robert Bosch Engineering and Business Solutions Private Limited, Bengaluru, Karnataka, India

1 Introduction

The novel coronavirus disease-2019 pandemic is the biggest public health epidemic faced by mankind. The virus has spread to every habited continent since its arrival in Asia in late 2019. Across all developed and developing nations, the cases are rising daily. The Exponential spread of the infection has led to a severe shortage of accurate testing kits as they can't be manufactured fast enough, creating panic amongst the citizens of several countries. This has resulted in the selling of bogus COVID-19 test kits and other fake vaccines to the public. The limited availability of accurate diagnostic test kits has resulted in an urgent need to focus on other methods for diagnosis. As COVID-19 attacks the epithelial cells which line our respiratory tract, we can use X-rays to examine the health of the lungs of a patient. Furthermore, provided that all major hospitals have access to X-ray imaging equipment, without the special test sets, X-rays could be used to monitor for COVID-19.

Currently, the only complication lies with the fact that the chest X-rays of COVID-19 patients have similar abnormalities with a Pneumonia Infected patient. Exploration is in progress to completely understand how COVID-19 pneumonia contrasts with different sorts of pneumonia. Data from these investigations can conceivably help find and facilitate our comprehension of how SARS-CoV-2 influences the lungs. So far, scientists have found that individuals with COVID-19 pneumonia were bound to have: (1) pneumonia that influences the two lungs rather than only one (2) lungs that had a trademark "ground-glass" appearance by means of CT check (3) abnormalities in some research tests, especially those evaluating liver

Novel AI and Data Science Advancements for Sustainability in the Era of COVID-19
https://doi.org/10.1016/B978-0-323-90054-6.00001-5

capacity. This clearly indicates that there is considerable room for the use of AI in diagnosing COVID-19 and differentiating it from viral pneumonia.

The Computer Vision groups across the globe have made huge efforts over the last decade and made many State of the Art models open to the public. These State-of-the-art models are conditioned on various data types and can be fine-tuned for certain typical tasks and purposes. For this analysis want to harness the capabilities and predictive power of pre-trained models to classify between COVID-19, non-COVID Pneumonia, and Normal.

2 Related work

Rousan, Elobeid, Karrar, et al. (2020) studied that chest CT scans and chest X-rays show characteristic radiographic findings in patients with COVID-19 pneumonia. The study aims at describing the chest X-ray findings and temporal radiographic changes in COVID-19 patients. The authors studied the X-rays of 88 COVID-19 confirmed patients. A total of 190 chest X-rays were obtained for the 88 patients. Thirty-one percent of the X-rays showed visible abnormalities. The most common finding on chest X-rays was peripheral ground glass opacities affecting the lower lobes. In the course of illness, the opacities progressed into consolidations peaking around 6–11 days. Thus they conclude that Chest X-ray can be used in the diagnosis and follow Yee and Raymond (2020) developed a pneumonia predictor using feature extraction from Transfer Learning. InceptionV3 was used as the feature extractor. K-Nearest Neighbor, Neural Network, and Support Vector Machines were used to classify the extracted features. The Neural Network model achieved the highest sensitivity of 84.1%, followed by Support vector machines and K-Nearest Neighbor Algorithm. Among all the classification models, the support vector machines model achieved the highest AUC of 93.1% for patients with COVID-19 pneumonia. Barstugan, Ozkaya, and Ozturk (2020b) used machine learning algorithms to classify between COVID-19 and non-COVID-19 images. The authors considered feature extraction techniques like gray-level size zone matrix and discrete wavelet transform. The extracted features were classified using a support vector machine and 2-, 5-, and 10-fold cross-validation. The authors achieved 99.68% of accuracy for the SVM trained using the GLSZM feature extraction method. Wang, Zha, Li, et al. (2020) proposed the use of deep learning to distinguish COVID-19 and other pneumonia types. The authors segmented and eliminated irrelevant areas. DenseNet121-FPN was implemented for lung segmentation, and COVID19-Net that had a

DenseNet-like structure, was proposed for classification purposes. The authors reported 0.87 ROC and 0.88 AUC scores for the validation sets. Kassani, Kassasni, Wesolowski, Schneider, and Deters (2020) introduced a feature extractor-based multi-method ensemble approach for computer-aided analysis of COVID-19 pneumonia. Six Machine learning algorithms were trained on the features extracted by CNNs to find the best combination of features and learners. Considering the high visual complexity of image data, proper deep feature extraction is considered a critical step in developing deep CNN models. The experimental results on the chest X-ray datasets showed that the features extracted by Dense-Net-121 and finally trained using the Bagging tree classifier generate the best predictions with 99.00% classification accuracy. Wang and Wong (2020) introduced COVID-Net, to detect COVID-19 from X-ray images of the chest. The COVID-Net architecture was designed from a mixture of 1×1 convolutions, depth-wise convolution, and residual modules to allow for deeper system design and prevent the issue of gradient disappearing. The dataset given was a mix of the COVID chest X-ray dataset provided by Cohen, Morrison, and Dao (2020b), and Kaggle chest X-ray images dataset (Kaggle, 2020) for multi-class classification of multi-class classification of normal vs bacterial vs COVID-19 infection dataset. The obtained accuracy of this study was 83.5%. Khan, Shah, and Bhat (2020) proposed CoroNet, to automatically detect COVID-19 from chest X-ray images. Coronet was built using the Xception architecture with ImageNet weights. CoroNet achieved an overall accuracy of 89%, precision of 93% and recall of 98.2% for 4-class cases being COVID-19, Viral and bacterial Pneumonia and Healthy. The same model achieved 95% accuracy for 3-class classification i.e., COVID-19, Pneumonia and Healthy. Chouhan et al. (2020) proposed a deep learning approach to classify pneumonia from chest X-rays using State of the art pre-trained models. They tested the performances of State of the art pre-trained models like AlexNet, DenseNet, and Inception V3 etc. to extract features. The extracted features were passed through individual classifiers and the predictions of individual architectures were obtained. An overall ensemble of all five pretrained models was observed to outperform all other models. Rajaraman et al. (2020) studied and found that performing Reiterative pruning and selecting the best pruned model improved the prediction accuracies and further helped minimize parameter numbers as redundant parameters which do not help improve the prediction performance are eliminated. Further they were able to better the performance by use of ensembles of pruned models. Awarding weights based on their predictions, the authors observed that the weighted averaging ensemble of the pruned models outperformed the other ensemble methods. Overall it was identified that combinations of iterative pruning of models and ensembles of models helped reduce prediction variance, model complexity. In this chapter, we evaluate four different approaches/hybrids using State of the art pre trained models so to achieve maximum Accuracy and have low False Negatives.

3 Modeling

3.1 PCA-feature ensembles

The baseline models are initialized with ImageNet weights and are used to extract the image features. To act as a feature extractor, the final softmax layer is removed. The features extracted for all the baseline models are combined and reduced by using PCA. The number of PCA components is selected so as to explain 90% of the total variance. These PCA features are finally passed through a dense 256 layer and a softmax for final predictions. The architecture of PCA-Feature Ensembles for the baseline model is depicted in Fig. 1.

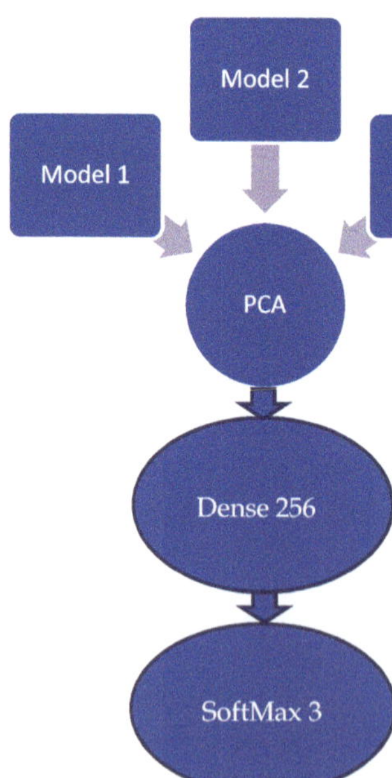

Fig. 1 Proposed feature ensemble for the baseline model.

3.2 Optimally weighted majority voting

This is a naïve but effective approach. The main Baseline models are individually assessed on the dataset and the probabilities prediction for all the classes are made. The prediction vector is a weighted average of the individual probabilities across all classes. The final prediction Y is the maximum probable class.

$$Y = \operatorname{argmax} \sum_{1}^{m} W_j \cdot P_{ij} \quad (1)$$

where W_j is the weight that can be assigned to the jth classifier.

The weights, W_j are calculated by a grid search so as to find best linear combination for most accuracy. Fig. 2 depicts the Weighted Majority Weighting ensemble.

3.3 Feature extraction

Feature extraction consists of using the representations learned by a previous network to extract distinguishing features from new samples. These features are then classified. The methodology involves (i) extracting the image features from the images (ii) The extracted features are then trained using a machine learning classification algorithm. The Feature extraction task is

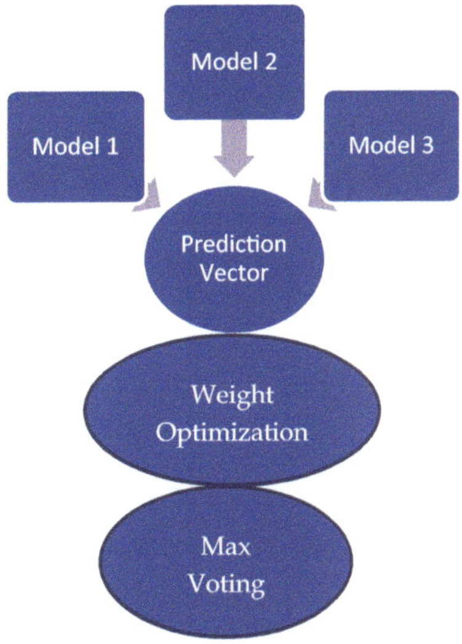

Fig. 2 Proposed weighted max voting model.

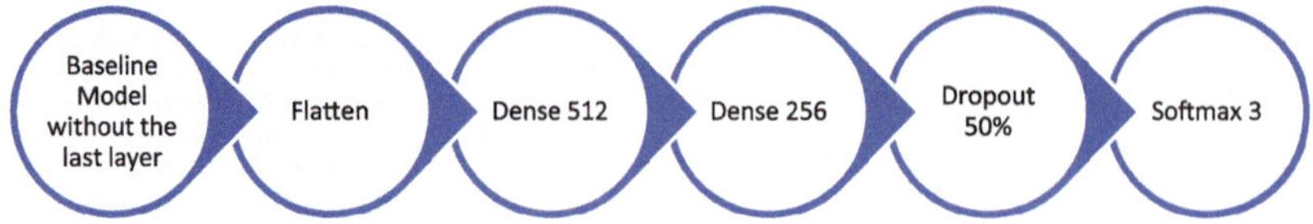

Fig. 3 Layer modification of baseline model.

performed using the various baseline models for comparison. For the process of classifying the features, we shall utilize the following three classification:

(i) Support Vector Machine (Cristianini, Shawe-Taylor, et al., 2000),
(ii) Bagging Classifier (Barstugan, Ozkaya, & Ozturk, 2020a) and
(iii) ADABOOST (Rosebrock, 2020) as previous works prove them to be constantly performing well for similar tasks.

3.4 Layer modification

The baseline networks are initialized with the weights from the ImageNet. The convolutional and max-pooling layers are frozen so that we don't modify their weights. The final softmax layer, mapping to 3 output classes, was replaced with 2 dense layers, 50% dropout layer, and softmax layer mapping to the X-ray labels. These layers were introduced to maximize baseline model classification accuracy during the transfer learning process. Once this is done, we would start retraining. In this way, we manage to take advantage of the feature extraction stage of our network and only tune the new additional layers to work better with our dataset.

Transfer learning by retraining the layers at all is not always a good idea. If the destination task is based on a small dataset that is very similar to the one the network was trained on, leaving the weights frozen and putting a classifier on top of the output probabilities is likely to be more useful, yielding largely similar results without risking overfitting. The architecture of layer modification for the baseline model is depicted in Fig. 3.

4 Experimental setup

4.1 Baseline models

In this section, we explain in brief about the selected pre-trained models which we will use as baseline models for our experiments.

4.1.1 VGG-16 (Simonyan & Zisserman, 2015)

VGG16 is a convolution neural net (CNN) network that was utilized to win ImageNet competition in 2014. Most remarkable thing about VGG16 is that as opposed to having countless hyper-parameter they concentrated on having convolution layers of 3 × 3 channel with a step 1 and consistently utilized same padding space and maxpool layer of 2 × 2 channel of stride 2. At last it has 2 fully associated layers with a softmax for final output. The 16 in VGG16 refers to it has 16 layers that have the weights. This system is a truly huge system and it has around 138 million parameters. VGG-16, although based off of AlexNet (Krizhevsky, Sutskever, & Hinton, 2017), it has the following key differences:

(a) It has replaced the large receptive fields of AlexNet's (11 × 11 with a stride of 4), with very small receptive fields (3 × 3 with a stride of 1). This introduces three ReLU units instead of just one, making the decision function to be more discriminative. Further this reduces the parameters (27 times the number of channels) instead of AlexNet's (49 times the number of channels).

(b) VGG-16 incorporates 1 × 1 convolutional layers to make the decision function more non-linear without changing the receptive fields.

(c) The small-size convolution filters allows VGG-16 to have a large number of weight layers; of course, more layers leads to improved performance.

4.1.2 ResNet 50 (He et al., 2016)

ResNet, short for Residual Networks is a classic neural network used as a backbone for many computer vision tasks. This model was the winner of ImageNet challenge in 2015. The key breakthrough with ResNet was it allowed training extremely deep neural networks with 150 + layers successfully. Prior to ResNet training very deep neural networks was difficult due to the problem of vanishing gradients. There are numerous variations of ResNet, for example same idea yet with a different number of layers. We have ResNet-50, ResNet-101, ResNet-110, ResNet-152 and so forth. The name ResNet followed by a two or more digit number basically suggests the ResNet design with a specific number of neural layers. ResNet-50 is one of the most compact and vibrant networks. The architecture of ResNet50 has 4 stages. The network can take the input image having height, width as multiples of 32 and 3 as channel width. Every ResNet architecture performs the initial convolution and max-pooling using 7 × 7 and 3 × 3 kernel sizes, respectively. Afterward, Stage 1 of the network starts and it has 3 Residual blocks containing 3 layers each. The size of kernels used to perform the convolution operation in all 3 layers of the block of stage 1 are 64, 64 and 128, respectively. The convolution operation in the Residual Block is performed with stride 2. Hence, the size of input will be reduced to half in terms of height and width but the channel width will be doubled. As we progress from one stage to another, the channel width is doubled, and the input size is reduced to half. For deeper networks like ResNet50, ResNet152, etc., bottleneck design is used. For each residual function F, 3 layers are stacked one over the other. The three layers are 1×1, 3×3, 1×1 convolutions. The 1×1 convolution layers are responsible for reducing and then restoring the dimensions. The 3×3 layer is left as a bottleneck with smaller input/output dimensions. Finally, the network has an Average Pooling layer followed by a fully connected layer having 1000 neurons.

4.1.3 Inception V3 (Szegedy et al., 2015)

Inception V1 was the winner of the ImageNet Competition 2014. It created the record lowest error rate at ImageNet dataset. The model is continuously improved so as to enhance the accuracy and decrease the complexity of the model. Inception V3 network stacks 11 inception modules where each module consists of pooling layers and convolutional filters with rectified linear units as activation function. The input of the model is two-dimensional images of 16 horizontal sections of the brain placed on 4 3 4 grids as produced by the preprocessing step. Three fully connected layers of size 1024, 512, and 3 are added to the final concatenation layer. A dropout with rate of 0.6 is applied before the fully connected layers as means of regularization. The model is pre-trained on ImageNet dataset and further fine-tuned with a batch size of 8 and learning rate of 0.0001. Inception V3 has the following changes compared to its previous models:

(a) Uses RMSProp optimizer instead of SGD l.
(b) Added Batch Normalization to the dense layer of the Auxiliary classifier.
(c) Uses of 7×7 factorized Convolution
(d) Label Smoothing Regularization: Regularizes the classifier by calculating the influence of label dropout during training. It penalizes and prevents the classifier from predicting very high probabilities for any single class. This improved the error rate by 0.2%.

We shortlisted these three architectures as our baseline as they have consistently shown good performance in regular image classification tasks and medical image classification tasks (Choi, 2015; Margeta, Criminisi, Lozoya, Lee, & Ayache, 2016; Tajbakhsh et al., 2016). Table 1 highlights the connection type, parameters and total floating-point operations in the three baseline models.

Table 1 Key points of the baseline models.

Model	Connection type	Parameters	Floating point operations
VGG-16	Fixed-kernel	138M	19.6 B
ResNet-50	Shortcut	23M	11 B
Inception V3	Wider-parallel	24M	2 B

4.2 Dataset

The dataset used comprises of labeled chest X-ray images of (i) COVID-19 infected (ii) Pneumonia infected and (iii) healthy people obtained from the following public sources;

(i) Kaggle Pneumonia dataset (1583 normal X-ray + 4273 pneumonia X-ray) (Shih et al., 2019).
(ii) Kaggle Covid-chest Dataset (150 COVID-19) (Kaggle, n.d.).
(iii) GitHub UCSD-AI4H/COVID-CT (288 COVID-19 X-ray) (GitHub, 2020).
(iv) SIIM.org (60 COVID-19 X-ray) (SIIM.org, n.d.).
(v) University of Montreal (684 COVID-19 X-ray) (Cohen, Morrison, & Dao, 2020a).

Fig. 4 shows the imbalance in the classes of X-ray images in the dataset. Pneumonia Infected X-rays constitute 61%, Healthy (Non-Pneumonia and Non-COVID-19) X-rays constitute 22% and the rest 17% are COVID-19 X-rays. Many classification algorithms have low predictive accuracy for the infrequent class. Thus we treat this imbalance by making use of data augmentation strategy to partially rectify this skew in data. Figs. 5–7 show random samples from the dataset for the 3 Classes. It can be noticed that for an untrained eye it's nearly impossible to predict and point out the opacities in chest X-ray.

4.3 Data augmentation

It must be noted that X-ray images are usually of high resolution i.e. usually 1024 pixels × 1024 pixels and are single-channel images and not RGB, unlike normal images. The most common data augmentation technique i.e., cropping of the images, will not be performed on X-ray images to ensure abnormalities within the images is not cropped out. Therefore we perform the following augmentation strategies:

(a) Flipping: We perform separate horizontal and vertical flips for each image dataset.

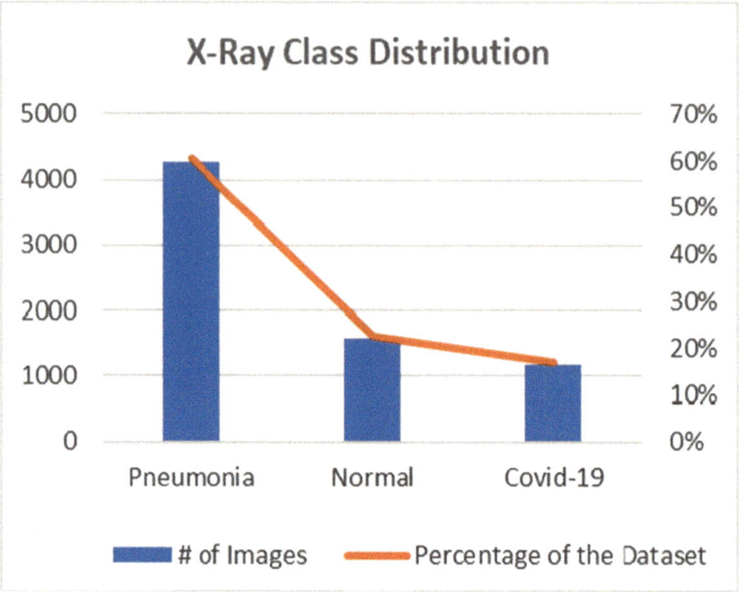

Fig. 4 Distribution of chest X-ray classes.

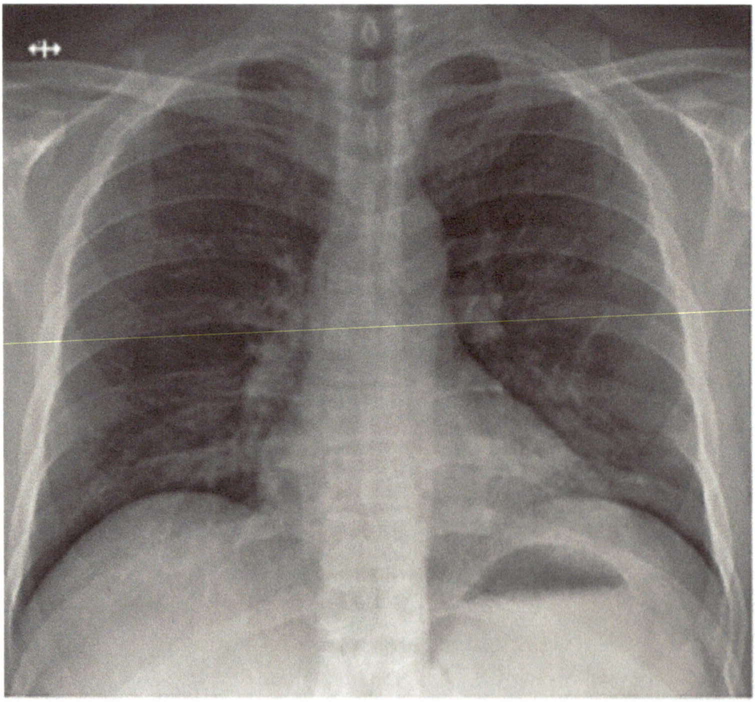

Fig. 5 COVID-19 positive chest X-ray.

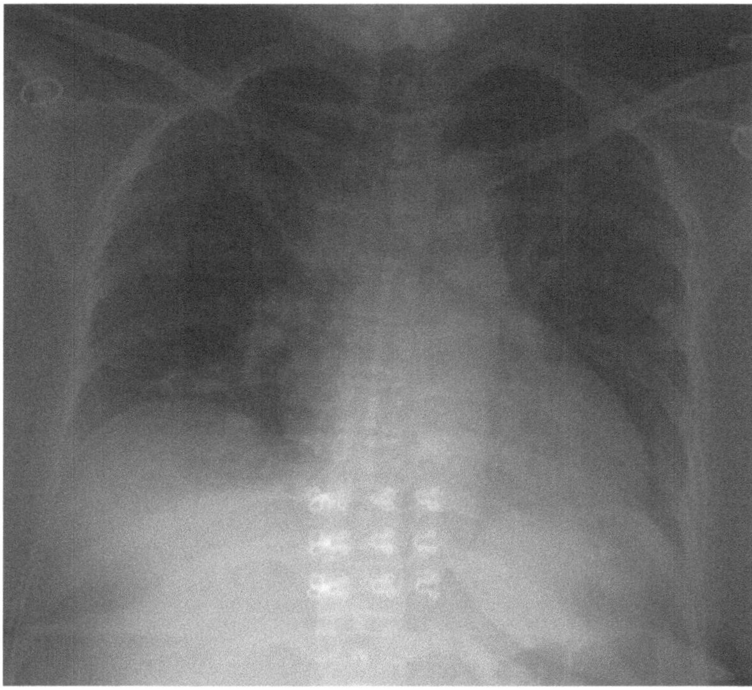

Fig. 6 Pneumonia (positive) chest X-ray.

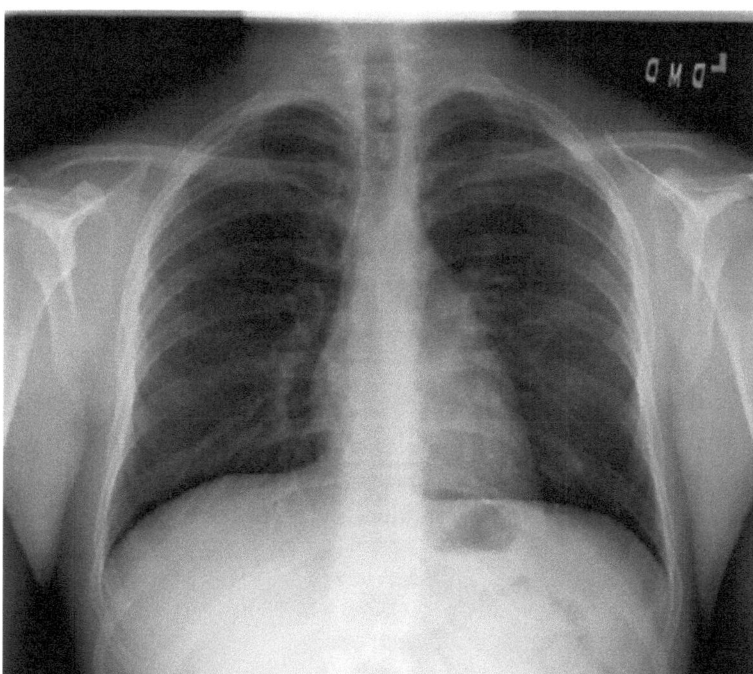

Fig. 7 Healthy chest X-ray.

(b) Rotation: Rotation of images is done using the following transformation,

$$A = \begin{pmatrix} \cos\theta & -\sin\theta \\ \sin\theta & \cos\theta \end{pmatrix}$$

where θ is between 10 and 90 degrees, is applied.

(c) Gaussian Noise: An array, A, is generated where each element in the array is a sample from a Gaussian distribution with $\mu = 0$ and with $\sigma 2$ in the range of [0.1, 0.9]. For each image X in the dataset, we obtain a noisy image, $X' = X + A$.

(d) Jitter: For each image in the dataset, we add a small amount of contrast (±1–5 intensity values).

(e) Power: For each image in the dataset, we take it to power. The power, p, is given by:

$$p = n \times r + 1$$

where n is a number taken from a Gaussian distribution with mean 0 and variance 1 while r is a number <1. Then, the augmented image, X_a, is given by,

$$X_a = \text{sign}(X) \times (|X|^p)$$

The sign and power are each taken elementwise.

(f) Gaussian Blur: A function defined by the variance between 0.1 and 0.9. ($r = 3\sigma$) is applied to blur the images

(g) Shearing: For each image in the dataset, the following transformation is done,

$$A = \begin{pmatrix} 1 & s \\ 0 & 1 \end{pmatrix}$$

S is the amount that image is to be sheared, and it is in the range of [0.1, 0.35].

4.4 Other preprocessing

The images vary in quality and dimension, ranging from 1215 × 759 pixels to 1024 × 1024 pixels due to multiple sources. To handle this issue we brought all the images to the size of 778 × 778 pixels to obtain a constant dimension for across all the input images.

4.5 Evaluation metrics

The Evaluation metrics are derived from the confusion matrix. Confusion Matrix is performance measure for most classification problems where output can be two or more classes. It is a table with four different combinations of predicted and actual values. (Confusion Matrix of N^2 combination can also be used to note the predictions v/s actual of all N classes). Table 2 shows a standard Confusion Matrix for a 2 Class case.

4.5.1 Accuracy

Classification accuracy is a naïve metric. It is the number of correct predictions made divided by the total number of predictions made. Accuracy in confusion metric terms is given by:

$$\text{Accuracy} = \frac{\text{True positive} + \text{True negative}}{\text{True positive} + \text{True negative} + \text{False positive} + \text{False negative}}$$

4.5.2 Precision

Precision can be thought of as a measure of a classifiers exactness. A low precision indicates a large number of False Positives. Precision in confusion metric terms is given by:

$$\text{Precision} = \frac{\text{True positive}}{\text{True positive} + \text{False positive}}$$

4.5.3 Recall

Recall calculates how many of the Actual Positives our model capture through labeling it as Positive (True Positive). Recall is the model metric we use to select our best model when there is a high cost associated with False Negative. Thus in Covid patient detection, If a Covid patient (Actual Positive) goes through the test and predicted as not sick

Table 2 Confusion matrix.

	Predicted	
Actual	**Negative**	**Positive**
Negative	True negative	False positive
Positive	False negative	True positive

(Predicted Negative). The cost associated with False Negative will be extremely high if the sickness is contagious. The recall in confusion metric terms is given by:

$$\text{Recall} = \frac{\text{True positive}}{\text{True positive} + \text{False negative}}$$

4.5.4 F-1 score

F1 Score is a good measure to use if we need to seek a balance between Precision and Recall and since there is an uneven class distribution of the COVID samples.

F1-Score in confusion metric terms is given by:

$$\text{F1} - \text{Score} = \frac{2 \times \text{Recall} \times \text{Precision}}{\text{Recall} + \text{Precision}}$$

4.6 Experimental details

The primary goal of our experiment is to utilize the power of altered transfer learning approaches to correctly diagnose COVID infection against Pneumonia infection and Normal-No infection using chest X-ray images. As discussed in Section 3, we have prepared 17 different models and studied them separately. For training, we used RMSProp optimizer and the cross-entropy loss function. The learning rate is started from the value of 0.001 and is reduced by 1 after every 5 epochs. The early stopping function takes care of the epoch number. The total images after augmentation processes and duplication removal was 211142 and 10% of this was held for testing.

5 Results and discussion

The pre-trained models are taken in their bare form as suggested by their respective papers for image classification without any alterations to get a benchmark. We have conducted the experiments using the methodology discussed in Section 3. Additional details to the methodology are as below:

(i) **Hybrid 1**: The feature ensemble model, the features are extracted individually from VGG-16, ResNet-50, and Inception V3 and combined to form a 4048 features. The new feature vector is reduced with PCA for 90% variance explained and passed through a dense layer and softmax.

(ii) **Hybrid 2**: The probability predictions of VGG-16, ResNet-50, and Inception V3 is passed through a weighted voting system to determine the final predictions. The weights are determined using a solver to ensure the three weights predict produce the best accuracy on the validation set.

(iii) **Hybrid 3:** Modified Architecture of the three Models are trained individually. This architecture allows us to take advantage of the feature extraction stage of our network and only tune the new additional layers to work better with our dataset.

(iv) **Hybrid 4:** The features extracted from VGG-16, ResNet-50, Inception V3 are passed separately through the three machine learning classifiers. This results in 3×3 combinations. This helps to identify which models produce the most distinguishable feature representation

The Accuracy, Precision, Recall and F1 score for all the hybrid models are reported in Table 3. It can be seen that VGG-16 although was the simplest of the 3 baseline models still outperforms the other 2 considerably for the Chest X-ray dataset. It achieves an F1-Score of 94.14. Fig. 8 shows the comparative results of the 3 baseline models.

In Hybrid 1: 1000 VGG-16 features, 2048 ResNet-50 features and 1000 Inception V3 features are individually extracted. These 4048 features are fed to a PCA to perform feature selection and develop a union feature set from them. The new feature set comprises of 1257 features explaining 92.64% of the total variance. The predictions after this enhanced feature set is passed through the dense and softmax layer produces a F1 score of 95.74 which outperforms all the 3 individual baseline models. This is mainly due to the fact that combined features has more representational capacity than the features from any single model. Additionally it can be noted the new feature set is smaller than features extracted from Resnet-50.

In Hybrid 2: The final voting weights were 0.43, 0.18, and 0.39 to attain the best F1 score. The Higher weight for VGG-16 can be explained from its performance in the baseline study. The combined prediction power of multiple models clearly outperform the baseline models as they achieve a F1 score of 96.19. This can be expected from an ensemble model as it helps utilize the power of individual model features. Fig. 9 shows the results of Hybrid 1 and Hybrid 2. It can be seen both have better performance than the Baseline models. This highlights the advantage and the power of model ensembles.

In Hybrid 3: The modified architecture has significantly improved the individual score of the models by an average of 9.8% as seen in Fig. 10. This is because of the extended architecture could take advantage of the feature

Table 3 Summary of results.

Type	Models	Accuracy (%)	Precision (%)	Recall (%)	F1-Score
Baseline	VGG-16	90.19	94.12	94.16	94.14
	ResNet-50	87.28	92.12	91.18	91.65
	Inception V3	82.22	85.12	85.24	85.18
Hybrid 1	Feature Ensembles	91.76	95.88	95.60	95.74
Hybrid 2	Majority Voting	92.19	96.33	96.06	96.19
Hybrid 3	VGG-16 Modified	**99.52**	**99.77**	**97.93**	**98.84**
	ResNet-50 Modified	97.75	95.80	94.83	95.31
	Inception V3 Modified	92.09	95.33	95.47	95.40
Hybrid 4	VGG-16 SVM	91.19	94.12	93.15	93.63
	VGG-16 Bagging	90.19	92.22	92.16	92.19
	VGG-16 AdaBoost	90.19	89.16	90.10	89.63
	ResNet-50 SVM	96.11	89.12	89.12	89.12
	ResNet-50 Bagging	95.12	95.12	95.07	95.09
	ResNet-50 AdaBoost	90.18	88.12	88.12	88.12
	Inception V3 SVM	84.29	85.12	85.20	85.16
	Inception V3 Bagging	**99.36**	**99.36**	**99.12**	**99.24**
	Inception V3 AdaBoost	80.12	85.00	86.12	85.56

Bold stands for best model.

extraction stage of our network and only tune the new additional layers to work better with our dataset. Modified VGG-16 achieved an accuracy of 99.52% It also is observed that the F-1 score of the Inception V3 model beats the ResNet-50 despite the accuracies of ResNet-50 is higher. Fig.10 depict the performance of Hybrid 3.

In Hybrid 4: The Bagging classifier performs best across all three models. The Inception V3-Bagging variant performs outstandingly with 99.36% accuracy (135/21114 misclassified). Fig.11 Compares the performance of various feature extractor and classifier combinations.

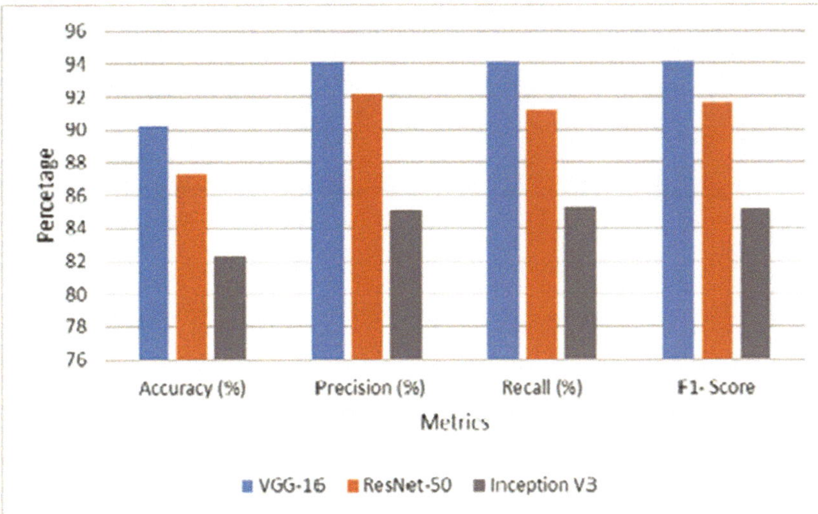

Fig. 8 Results of baseline models.

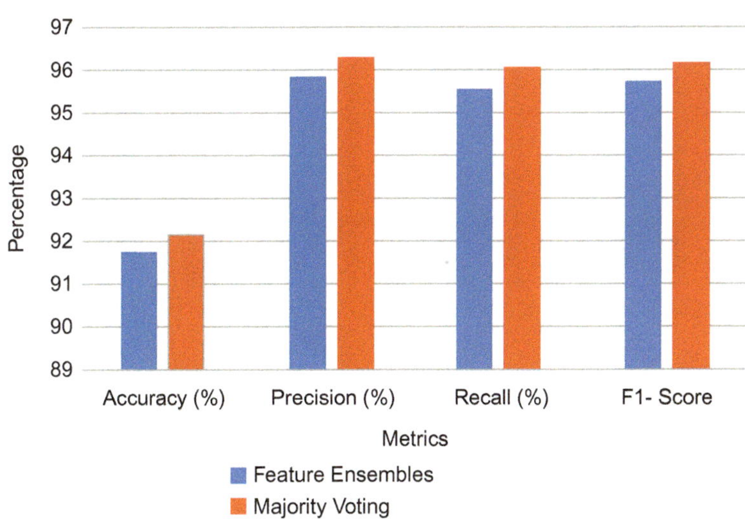

Fig. 9 Results of Hybrid 1(feature ensemble) and Hybrid 2 (max voting).

Overall best performer is the Hybrid 3-VGG-16 with modified layers with an accuracy of 99.52% and F1-score of 98.84. The confusion matrix of the same is shown in Table 4. It can be seen that out of 21,114 images, only 101 were misclassified. The achieved accuracy to 99.52% is far higher than any testing kit available in the market. Another breakthrough is the fact

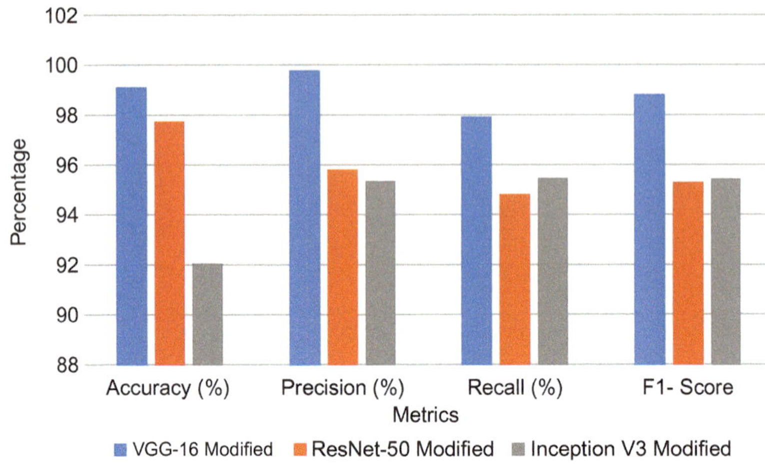

Fig. 10 Hybrid 3 model.

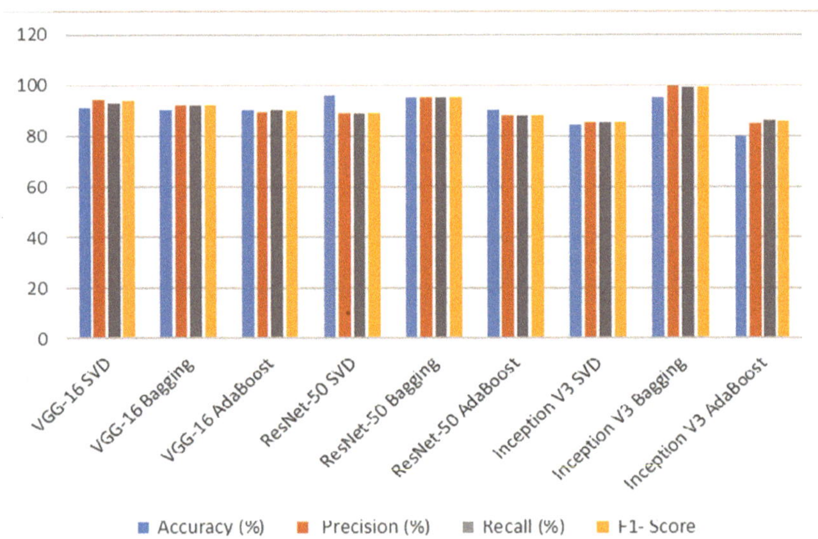

Fig. 11 Hybrid 4 model.

Table 4 Confusion matrix modified VGG-16.

	True		
Predicted	**COVID-19**	**Pneumonia**	**Normal**
COVID-19	3523	25	0
Pneumonia	76	12,765	0
Normal	0	0	4725

that there were ZERO cases where the infection was recorded as normal (False Negative Normal).

6 Conclusions

In this chapter, we propose a quick diagnostic tool using ensemble/hybrid approaches to classify COVID-19 and pneumonia from chest X-ray images using pre-trained models. We explored 4 possible hybrid methods incorporating pre-trained architectures like Inception V3, VGG-16, and ResNet18 trained on the ImageNet dataset. We used the 3 architectures for Feature extraction and ensemble prediction. We found that the modified VGG-16 and the Inception v3 + Bagging achieved accuracies of 99.52% and 99.36%, respectively. For future study, we propose increasing the dataset size and using hand-crafted features. Our findings support the notion that deep learning—AI approaches can be used to improve and ease the diagnostic process and improve disease management.

References

Barstugan, M., Ozkaya, U., & Ozturk, S. (2020a). *Coronavirus (COVID-19) classification using CT images by machine learning methods*. arXiv preprint arXiv:2003.09424.

Barstugan, M., Ozkaya, U., & Ozturk, S. (2020b). *Coronavirus (COVID-19) classification using CT images by machine learning methods*. ArXiv https://arXiv:2003.09424.

Choi, S. (2015). X-ray image body part clustering using deep convolutional neural network: SNUMedinfo at imageCLEF 2015 medical clustering task. In *Proc. workshop CLEF 2015 working notes*.

Chouhan, V., Singh, S. K., Khamparia, A., Gupta, D., Tiwari, P., Moreira, C., et al. (2020). A novel transfer learning based approach for pneumonia detection in chest X-ray images. *Applied Sciences*, *10*(2), 559.

Cohen, J. P., Morrison, P., & Dao, L. (2020a). *COVID-19 image data collection*. [Online] Available: https://arxiv.org/abs/2003.11597.

Cohen, J. P., Morrison, P., & Dao, L. (2020b). *Covid-19 image data collection*. arXiv 2003.11597 https://github.com/ieee8023/covid-chestxray-dataset.

Cristianini, N., Shawe-Taylor, J., et al. (2000). *An introduction to support vector machines and other kernel-based learning methods*. Cambridge University Press.

GitHub. (2020). https://github.com/ieee8023/covid-chestxray-dataset. from April 16, 2020-May 5, 2020.

He, K., et al. (2016). Deep residual learning for image recognition. In *Proceedings of the IEEE conference on computer vision and pattern recognition*.

Kaggle. (2020). *Kaggle's chest X-ray images (pneumonia) dataset*. https://www.kaggle.com/paultimothymooney/chest-xray-pneumonia.

Kaggle. Available: https://www.kaggle.com/andrewmvd/convid19-xrays.

Kassani, S. H., Kassasni, P. H., Wesolowski, M. J., Schneider, K. A., & Deters, R. (2020). *Automatic detection of coronavirus disease (covid-19) in X-ray and CT images: A machine learning-based approach*. arXiv preprint arXiv: 2004.10641.

Khan, A. I., Shah, J. L., & Bhat, M. M. (2020). CoroNet: A deep neural network for detection and diagnosis of COVID-19 from chest X-ray images. *Computer Methods and Programs in Biomedicine, 196*, 105581. Crossref. Web.
Krizhevsky, A., Sutskever, I., & Hinton, G. E. (2017). Imagenet classification with deep convolutional neural networks. *Communications of the ACM, 60*(6), 84–90.
Margeta, J., Criminisi, A., Lozoya, R. C., Lee, D., & Ayache, N. (2016). Finetuned convolutional neural nets for cardiac MRI acquisition plane recognition. *Computer Methods in Biomechanics and Biomedical Engineering Imaging and Visualization*. https://doi.org/10.1080/21681163.2015.1061448.
Rajaraman, S., Siegelman, J., Alderson, P. O., Folio, L. S., Folio, L. R., & Antani, S. K. (2020). *Iteratively pruned deep learning ensembles for covid-19 detection in chest X-rays*. arXiv preprint arXiv:2004.08379.
Rosebrock, A. (2020). *Detecting COVID-19 in X-ray images with Keras, tensor flow, and deep learning*. https://www.pyimagesearch.com/2020/03/16/detecting-covid-19-in-x-ray-images-with-keras-tensorflow-and-deep-learning/.
Rousan, L. A., Elobeid, E., Karrar, M., et al. (2020). Chest x-ray findings and temporal lung changes in patients with COVID-19 pneumonia. *BMC Pulmonary Medicine, 20*, 245. https://doi.org/10.1186/s12890-020-01286-5.
Shih, G., et al. (2019). Augmenting the National Institutes of Health chest radiograph dataset with expert annotations of possible pneumonia. *Radiology. Artificial Intelligence, 1*(1), 1–5.
SIIM.org. https://SIIM.org/covid-chestxray-dataset.
Simonyan, K., & Zisserman, A. (2015). Very deep convolutional networks for large-scale image recognition. In *International conference on learning representations*.
Szegedy, C., et al. (Jun. 2015). Going deeper with convolutions. In *Proc. IEEE conf. comput. vis. pattern recognit* (pp. 1–9).
Tajbakhsh, N., et al. (2016). Convolutional neural networks for medical image analysis: Full training or fine-tuning? *IEEE Transactions on Medical Imaging, 35*(5), 1299–1312.
Wang, L., & Wong, A. (2020). *COVID-net: A tailored deep convolutional neural network design for detection of COVID-19 cases from chest radiography images*. arXiv preprint arXiv:2003.09871.
Wang, S., Zha, Y., Li, W., et al. (2020). A fully automatic deep learning system for COVID-19 diagnostic and prognostic analysis. *European Respiratory Journal, 56*(2), 2000775.
Yee, S. L. K., & Raymond, W. J. K. (2020). Pneumonia diagnosis using chest X-ray images and machine learning. In *Proceedings of the 2020 10th international conference on biomedical engineering and technology (ICBET 2020)* (pp. 101–105). New York, NY: Association for Computing Machinery. https://doi.org/10.1145/3397391.3397412.

CHAPTER 2

Investigation of COVID-19 and scientific analysis big data analytics with the help of machine learning

Victor Chang[c], Mohamed Aleem Ali[a], and Alamgir Hossain[b]
[a]Cybersecurity, Information Systems and AI Research Group, School of Computing, Engineering and Digital Technologies, Teesside University, Middlesbrough, United Kingdom
[b]Academic Affairs & Research, Cambodia University of Technology and Science (CamTECH), Phnom Penh, Cambodia
[c]Department of Operations and Information Management, Aston Business School, Aston University, Birmingham, United Kingdom

1 Introduction and background

This project is based on the investigation process of COVID-19 and scientific analysis, big data with the assistance of machine learning. The extended COVID-19 is a global epidemic that has produced an exponentially rising data value that may be attached to enhance the considerate research of big data management. Additionally, to demonstrate the importance of policymakers, scholars, and practitioners for a deeper and better understanding of the assortment of investigative tools that might be exploited to respond better and anticipate unexpected risks and events. Deal with the global pandemic, scientists, clinicians, and WHO (World Health Organization) are investigating or searching the new technologies to analyze an infected person in different stages, to control the virus spread, develop the vaccine for a remedial infected person. Current investigation recognized that artificial intelligence and machine learning are capable technology engaged with the different providers of health care as they consequence the better power of processing and obtain a better result with accurate human care work. This project will explain the solving problems of COVID-19 with the benefit of machine learning. The objectives will describe the methods, techniques, and accurate process of data analysis with a different algorithm utilized in machine learning. The main focus of this project is on the pandemic of COVID-19 and how the technologies of machine learning are currently engaged to resolve the disputes during the situation of COVID-19. See Fig. 1.

https://doi.org/10.1016/B978-0-323-90054-6.00007-6

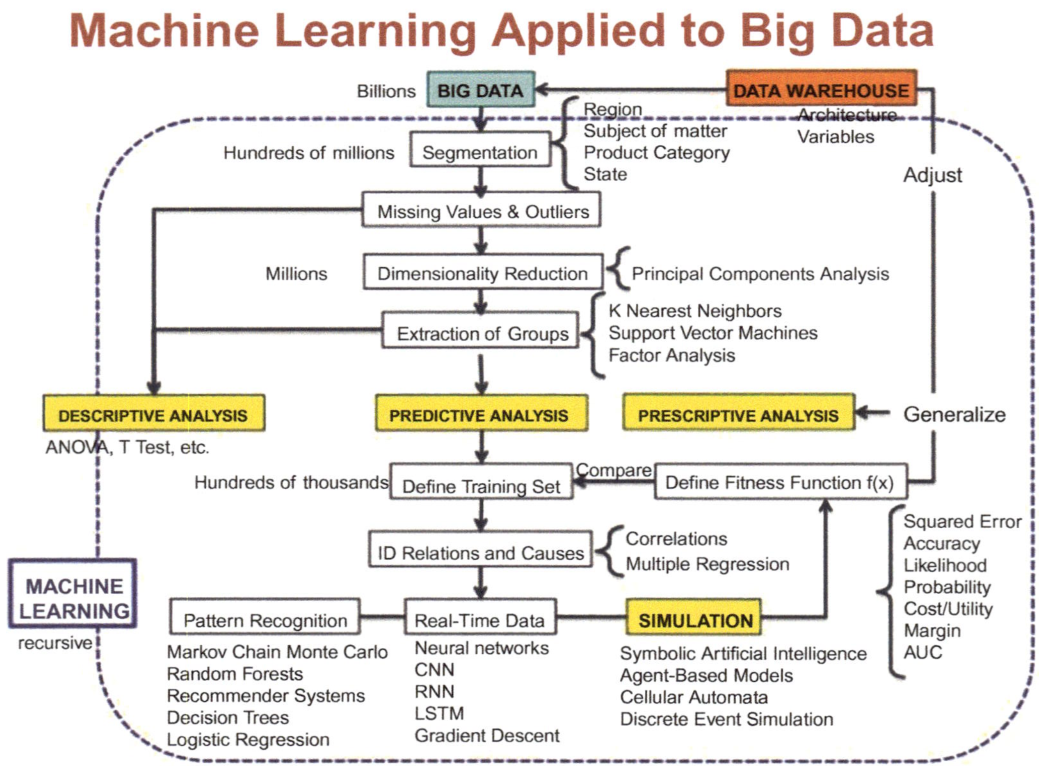

Fig. 1 Machine learning in big data. *(https://analytixon.com/images/.)*

"2019-nCoV" or "the novel coronavirus" or "COVID-19" as given the same by the world health organization (WHO), is the responsible organization regarding the present outburst of pneumonia, the commencement of which was in December 2019 in the urban of Wuhan, China. This virus is pathogenic. The reservoir of occurrence of this virus are bats, but the host that served as an intermediary has still been undetected. Though, there is ongoing work in china in three very significant areas regarding the advising on the awareness of the pathogenic origination of the outbreak. This involves inquiries of the cases who are showing some symptoms, which were occurring in the city of Wuhan around 2019, December, collection of ecological samples from the market of wholesale seafood, and also additional markets within such areas, the collection regarding the reports made on the detailing of the origination, the kinds of wildlife species which were marketed in those markets, and what are the destination of such animals once there has been closing of the markets. Mostly the gastrointestinal and the respiratory tract infections are caused by the coronavirus and they are categorized and four significant kinds: Delta coronavirus, Gamma coronavirus, and Alpha coronavirus and Beta coronavirus. Infection is caused to birds from the first two kinds of the coronavirus, while mammals are infected from the last two. When it comes to human beings, coronavirus has been formally categorized into six kinds. The Middle East Respiratory Syndrome coronavirus (MERS-CoV), HCoV-OC43, HCoVHKU1, Severe Acute Respiratory Syndrome coronavirus (SARS-CoV), which is the kind of the Beta coronavirus, HCoV229E as well as HCoV-NL63, which are the associate of the Alpha coronavirus. Coronavirus was not considered a global concern until 2003 when the pandemic of SARS arrived, which first-order continued in the year 2012 by MERS, and then with the most recent cases and outbreaks by the COVID-19. All these three viruses are highly pathogenic. Their spreading is from bats to palm civets, and then in the end to the humans. Suppose there will be unsafe touching between the infected person and the person who can be infected, COVID-19 splendid given by dust particles and the fomites. Although based on the empirical evidence, transmission through the air is not recorded for coronavirus, and it is not considered a very important transmission engine. However, there is a possibility that if such particles, which are aerosol-generating, will be carried out in the medical facilities, they can become an important mode of transmission. In some patients, it was observed that it could also be spread fecal spreading. Also, the fecal and oral routes are not quite identified much significantly. Their functions and the importance of spreading COVID-19 are an ongoing process and still need to be determined. In the second week of April 2020, approximately 1,873,858 laboratory-confirmed cases were recorded. After further studying the age groups, it was determinable that a maximum number of confirmed cases bar between the age groups of 30–69. In all the recorded cases, approximately 21.6% of the cases were found in the working class or employees by profession and farmers. The 2019 novel coronavirus, or acute respiratory syndrome known nowadays, is rapidly spreading from China to the rest of the world. Around 95,000 cases of the disease in 2019 and 34,000 death have been recorded. This is fortunate that the children are not affected by the virus with no deaths. But the cause of the future is unknown.

However, currently there is too much to be concerned about in relation to the latest coronavirus. However, it can be said that the humans were infected because of transmission through animals. Therefore, it becomes necessary to determine particular animals and various other sources, the line of the broadcast, the sequence of incubation, characteristics of the susceptible public, and survival rate. However, there is not much clinical knowledge of the COVID-19 and other details regarding it. Like the outbreak curve, the spectroscopy of its virility, the pathogenesis of its dissemination, the observations of the autopsy, and any types of clinical responses. COVID-19 is a lot different than other viruses that have emerged in history. The reach of COVID-19 has been to almost 150 nations and the country of China. The outbreak of this virus was professed as a worldwide epidemic, according to the world health organization. In a very short span of time and only in the second week of April 2020, the total number of cases of COVID-19 exceeded 1,873,858 and the deaths were recorded of 1,160,450 worldwide. The global epicenter of the coronavirus was the United States of America. Though the outbreak started in China, more than one-third of the coronavirus cases were found outside the country of China. It was very difficult to establish any kind of connection between the pandemics that occurred in the past and this one, for example, swine flu, bird flu and more so. Since it was so new and no vaccine could be made in such a short period, the only way left to fight this virus was some precautions for its prevention. Some kinds of preventive measures and policies were advised and enforced by the world health organization. Since this epidemic is continuous and increasing rapidly, some precautions are necessary to be adopted, such that the risks of its spreading could be significantly reduced and its transmission chain could be broken. The advice given by the world health organization is:

Regular washing of hands with any hand wash that is alcohol-based or soap and water is helpful.

Proper distance maintenance.

There is no touching of nose, head, and ears.

It requires a Proper covering of the nose and mouth to prevent the spread of viruses in the air.

In case any person has fatigue or faces any trouble in breathing, it should be taken seriously and provided with early medical attention.

Take proper precautions, like isolation, if recently any place has been visited by you where there are high chances of spreading coronavirus.

Since a progressive "datafication" has been witnessed in the social life. There has been monitoring of human activities as well as the interactions of humans with the environment, with effectiveness such that a huge amount of digital footprints can be generated. Big data examination is akin to a very complicated procedure of examining data to uncover important information. For example, correlations, the customers' preferences, many hidden patterns, and many kinds of market trends will be helpful for the organization in making correct and informed decisions.

On a wider scale, the technologies and techniques of data analytics are very significant for the organization since they provide a way in which sets of data could be analyzed and new information could be gathered. Basic questions queries regarding the performance and operations of the business might be answered with business intelligence support. Big data analytics can also be referred to as advanced analytics, which includes complicated applications with traits like predictive models, the algorithms of statistics and analytics system-powered what-if analysis.

Since the collection, processing, and cleaning of the data are done, the data will be analyzed with analytics software. This involves the tools for the mining of data, which sifts through the sets of data in search of establishing or determining any kinds of patterns or relationships. Text mining, predictive analysis-helps build software that can forecast the customer's behaviors and other developments related to the future. Artificial intelligence and visualization tools for data science and deep learning can also be termed the advanced offshoot of data-oriented machine learning. Some examples of key big data analytics tools which are put into use for the enabling of big data analytics processes involve:

- Hadoop: is an open basis outline for the storage and dispensation of big data sets. Hadoop can handle huge amounts of organized as well as formless data. According to Hadoop, the advantages include the following characteristics.

Performance, scalability, data locality, support for multiple languages, open source, schema independent, fault tolerant, share-nothing architecture, cost-effectiveness, high-throughout and low latency, with explanations as follows.

- Predictive analysis: predictive analysis hardware as well as software process a huge amount of complicated data and make use of statistical algorithms and machine learning techniques so that predictions regarding the outcomes of future events could be made. Organizations use predictive analytical tools for the detection of fraud, marketing, and risk assessment.
- Stream analytics tools are used to filtrate, aggregate, and analyze big data that may stow in various arrangements or on different stages.
- Distributed storage data: the replication of which is done on non-relational databases. Its use could be made to a degree in contradiction of the bulge failures, which are independent, data that is not apt or corrupted.
- NoSQL databases: these are useful when huge sets of distributed data are used, and these are non-relational data management systems. A fixed scheme is not used or required by them, and this feature makes them ideal for the data, which is raw and unstructured.
- Data Lake is like a repository that helps store enormous amounts of data unruffled by various sources. The data which the data warehouses typically store is done by using predefined schemas.
- Big data mining tools: the enabled business in mining huge amounts of data is structured and unstructured data.
- In-memory data fabric: it has a distribution of huge amounts of data across the systems of mineral resources. It helps in hiding low latency for the data access as well as processing.
- Data integration software has enabled the big data to be rationalized on various stands, including Hadoop, Amazon EMR, Mongo DB, and apache.
- Spark: it is a cluster of open-source computing frameworks used for the batch and the streaming of data processing.
- Data quality software: cleanses as well as provides enrichment of the big data sets.

Machine knowledge is a way of data analysis due to which analytical model structure is computerized. It is like a division of artificial intelligence based on the idea start a lot could be learned by the systems from the data, like identifying patterns and making decisions with the minimum kind of human interruption. Due to the revolution of new computing technologies, machine learning in the present scenario does not write machine learning in the past. Its origination was recognizing pattern yes theory that the computers could learn without programming for the performance of those particular tasks, so many researchers were interested in the concept of artificial intelligence wanted to observe is learning from data could be done by computers.

The repetitive aspect in machine learning is very significant since models have to face the risk of exposure to new data. They have the ability to adapt independently. Learning is done from the competitions that were previously performed to produce reliable, repeatable decisions and results. This is a science that is not new, but it has been considered change and has definitely gained new and fresh momentum.

Recommendations regarding online offers like those from Amazon and Netflix can be said to be machine learning applications put into use in everyday life. Understanding and knowing what is being said by the customers about you and your organization on media platforms like Twitter is a classic example of machine learning combined with linguist rule creation. Detection of fraud is one of the most obvious and significant used in today's world.

The reasons for researching interest in machine learning are that some factors obligate Bayesian analysis and data mining much more famous than they were constantly. There are also other concerns, such as increasing volumes and the cost of computational processing. Users need a much lower and comparatively powerful and affordable data storage. In summary, it has become reasonable to ghastly and repeatedly yields the models that can examine bigger, more complicated data. The delivery will be much faster and the results will be more accurate. And this all will be done even if the scale of data is huge. Once precise models are built, the organization automatically has a much better chance of identifying profitable opportunities and avoiding unknown risks.

Many industries whose daily work is to go through huge quantities of data have understood the importance of machine learning technology. Through getting detailed visions from such usage data, most often in the actual time, establishments are now capable of working more efficiently and gaining an advantage over their competitors. Organizations provide financial services such as banks and many government agencies like public safety, healthcare industries, so many websites, organizations that are engaged in building new energy sources, and transportation industries—they can generate abundant data sources.

Machine learning depends upon CT analysis, which is also suggested as the promising screening tool for COVID-19. There is one more PCR test that is needed to do on the COVID-19 patient—the result to be cautiously understood. During the pandemic, the studies done of hamper inherently by the disease occurrence and the nature with the selected of the members, evaluation of CT. See Fig. 2.

The procedure also needs to be skilled on the variety of diseases, counting early-stage cases and asymptomatic. The elucidation of CT machine learning will be functional in real-world data with poise. Analytics of the big data calculator mortality risk from the COVID-19: the pandemic has warned the people, health officials searching continuously for the different tool to help the resources and give guidance in decision-making. There is a squad from John Hopkins school of public health purchase Big data analytics to build COVID-19 risk calculator, which well informed the public health policy around the resources of prevention like face mask (N95).

Artificial intelligence shows that the vaccine of the COVID-19 is less effective. With the use of authorization, the FDA recently granted the emergency for the new vaccines of the COVID-19 Hindi looking forward to the beginning to the end of the pandemic. The researcher news the tool to investigate the vaccine, which is similar to the COVID-19 vaccine and found it less effective in Asian ancestry or black people. The result highlights the ethnic disparities and stark racial, which have been highlighted consistently throughout the pandemic. Data set of the Google map will be freely available to a researcher: the new ways parked with the name of the COVID-19 of sharing the data and access in the healthcare. Google announced in late March that it would permit access to the critical coronavirus data to the researcher through its COVID-19 public data set program, which has the objective to boost the solution during the global pandemic.

Using machine learning to discover the COVID-19 solution by the data: India is at a pandemic stage. The researcher is working to find the therapies for COVID-19 which is potential by using the machine learning tool and artificial intelligence. To graduates from the institute of data science at Columbia University build a startup called EVQLV, which made the algorithm capable of screening, optimizing and computationally generating millions of antibodies. Find the COVID-19 patients who are at high risk. In

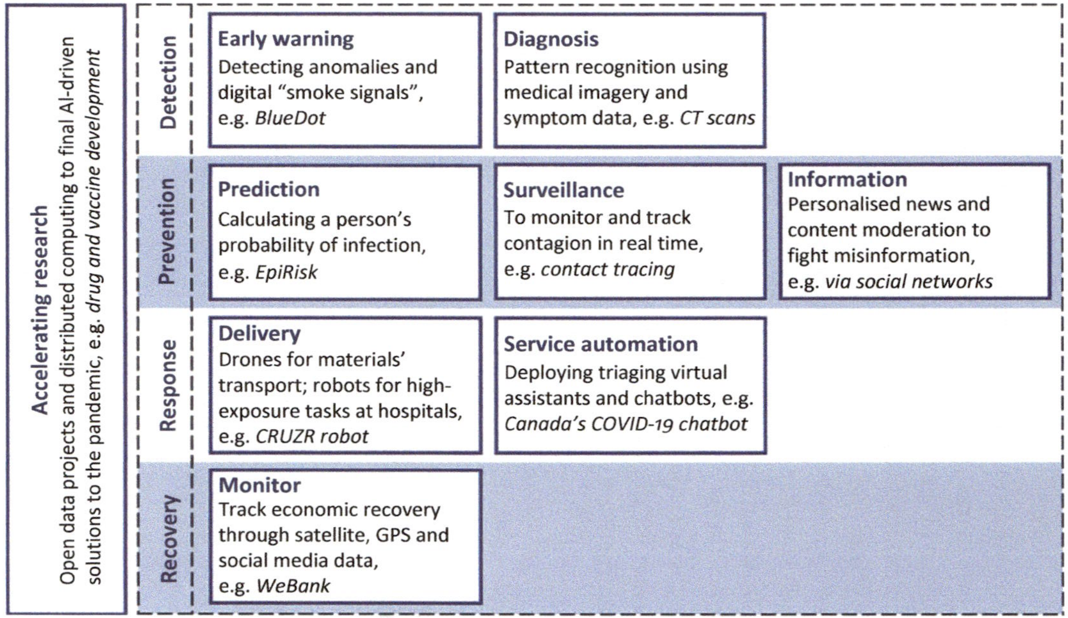

Fig. 2 Use of machine learning in COVID-19. *(Data from Nirmala, A. P., & More, S. (2020, December). Role of Artificial Intelligence in fighting against COVID-19. In* 2020 IEEE International Conference on Advances and Developments in Electrical and Electronics Engineering (ICADEE) *(pp. 1-5). IEEE.)*

The beginning, the virus spreading in the whole us, the population was at a higher risk while spreading this virus.

The virus has impacted the people or the elderly with good condition and affects the population who are the minorities and individuals of the lower status.

Accepting the COVID-19 as a big data analytic issue: the data has been considered as the midpoint of the pandemic Investigation effort. The extent of the coronavirus has resulted in its intricate nature and requires assessors and analysis of the massive amount of information. The particular interview has determined that there are different ways in which resources and development can use machine learning, natural language, artificial intelligence, and dispensation to track, understand, and contain coronavirus. The institute has been providing the government entities industries and Research Organization exist and innovation in artificial intelligence tools. Data platform helps track Hospital bed capacity during the outbreak of COVID-19: the spreading of COVID-19 has meant that hospitals are required to prepare for the poorest. In a particular system that has previously been tense, there are probable waves that result in highly contagious patients and can result in disaster translation. With the usage of Big Data Analytics, the organization can monitor and track the use of critical resources. Occasionally, the speed and accuracy of machine learning algorithms are to be stated on the foundation of the presentation in scenarios that are not capable of reflecting the actual clinical practice. Sometimes, the comparisons made between the algorithms and the preferences of humans are very imbalanced. In many cases, a computer has been trained to detect the particular abnormality related to parenchymal disease related to COVID-19. In contrast, any abnormality which is related to the finding of the pulmonary nodules becomes the responsibility of the radiologist to detect the same. There is no doubt that many exciting opportunities were offered by the COVID-19 pandemic regarding artificial intelligence research. With all the understanding which could be developed regarding the coronavirus, it is very easily determinable that this virus progresses when it reaches different stages. Regarding the previous pandemics that have occurred and their in depth-details, it can be said that this pandemic is not very different from the others. With the help of the correct collaboration between the clinical and machine learning know-how, the present public health crisis strength marks the starting of the decade when artificial intelligence regarding health care will deliver according to its promises of transformative clinical impacts.

2 Literature review

As per the words of Sheng, Amankwah-Amoah, Khan, and Wang (2020), COVID-19 has become a worldwide health disaster, approximately 16 million persons are infected by this virus and more than 666,000 deaths are faced till the 29th of July 2020 all over the world. This resulted in excessive demand for resources in several countries to stop the spread of the COVID-19 pandemic. COVID-19 disease confirmed its first case in December 2019 in China. The health crisis is demanding effective vaccines and drugs for treatment processes and mitigating the spread of infection. The majority of the contaminated areas follow lockdown policies, quarantine, and social distancing measures because of the limited validated therapeutics to reduce the spread of infection. These measures are proven helpful in slowing down the spread of infection, but it is not a permanent solution to eliminate the infection. The fear of spreading infection into the second wave was increasing. To limit the second potential outbreak, several measures like identifying hotspots, contact tracing are required.

Inconsistency has been observed in the symptoms of the COVID-19; the symptoms range from flu to fulminant pneumonia and serious respiratory distress syndrome. Various techniques such as big data analytics and machine learning investigate the information regarding the COVID-19 and scientific analysis. Machine learning is used to harness, predict, understand, and respond to be taken for future events.

As per the words of Lalmuanawma, Hussain, and Chhakchhuak (2020), numerous health care experts, clinicians, and scientists worldwide are working on researching new and suitable technologies for supporting the handling of the COVID-19 pandemic. Previous application of Machine Learning and Artificial Intelligence on this epidemic situation encouraged the investigators by providing a novel approach for battling against the coronavirus outbreak. Artificial intelligence, besides machine learning, provides a vital approach in screening, forecasting, predicting, and developing a drug for SARS-CoV-2 and contact tracing. These approaches helped the researchers by enlarging various angles towards the research. WHO and co-operating clinicians and several national authorities all over the world are fighting the pandemic? WHO avowed the COVID-19 situation as a global well-being emergency. Several researchers and scientists are searching for technology for screening the virus-infected patients, finding an appropriate clinical trial, developing a cure vaccine, controlling virus spread, and tracing the contact of virus-infected patients. Several healthcare providers hire machine learning and artificial intelligence because they result in reliable, speedy processing power, better scaling, and outperforming humans' in particular healthcare tasks. Most of the clinician and health care industries employed several AI and ML technology for handling the pandemic and addressing the challenges at the time of the outbreak.

AI in the industry of medicine is used for supporting the decisions of the clinician. AI is used for selecting a supplementary medication for investigating about proposed SARS-CoV-2. There are several technologies and methodologies are used for classifying and addressing the classical methods based on statistics to advanced and modern ML and AI algorithms. These technologies are promising improvements in predicting, screening, forecasting, and tracing contact and developing drugs. They help predict incidents and the number of deaths due to pandemics. They are used to perform predictive analysis, diagnostic and descriptive analytics, and build complex models for enlarging the datasets to predict future outcomes (Corsi, de Souza, Pagani, & Kovaleski, 2020).

In the early period of the spread and outbreak of COVID-19, it was necessary to adopt scientific data analysis methods for making effective decisions regarding the prevention of the COVID-19 situation. With the increasing popularity of technology on the internet, people around the globe can easily interact, communicate, learn, shop and others. See Fig. 3 for details.

Hence, using appropriate technology for analyzing big amounts of data associated with information networks, accurate predictions can be made on outbreak and blowout of COVID-19, way and cause of spread, and the next stage of spreading can formulate policy measures accumulated with it. Controlling and preventing these types of situations are classified in different stages: monitoring of the epidemic situation, early warning, and prevention of disaster, disaster recovery, and relief.

Machine learning and AI technology are useful in speeding up processing power, scaling up, and outperforming humans in specific healthcare tasks. Various healthcare clinicians and industries employ the technology of ML and AL for tackling up the situation of the COVID-19 epidemic for addressing challenges during the time of the outbreak. Even if AI cannot replace human interaction in healthcare, it can support decisions (Syeda et al., 2021).

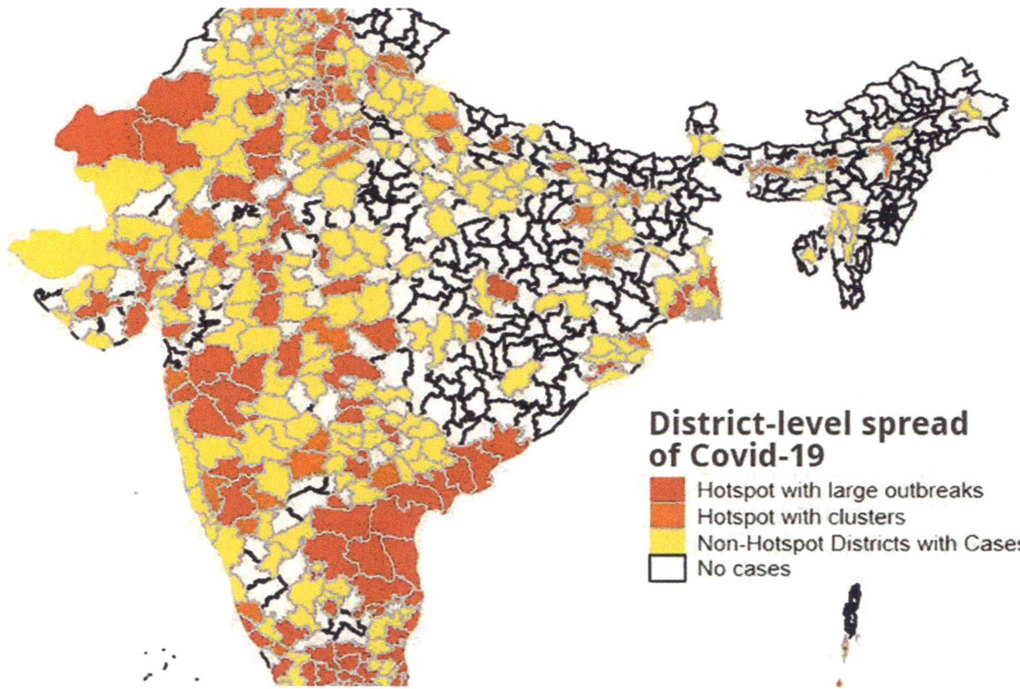

Fig. 3 COVID-19 statistics. *(Source india.gov.in; https://scroll.in/author/18857.)*

3 COVID-19 pandemic in the new era of big data analytics: Methodological innovations and future research directions

As per Jia, Guo, Wang, and Barnes (2020), the outbreak of the COVID-19 has generated the epidemic globally with an extraordinary volume of data and exponential amounting, which many enhance the empathy of the data administration research. The amount of data can be very huge and a symbolized need among the toppers policymakers and practitioners deeper and better understanding of the range for the analytical tool that could be used to respond better unanticipated to search unexpected Black swan risk and event. See Fig. 4.

Big data is specified by volume, which has great variety, complexity and velocity. The big data benefit does not seem to result from the size of data but only appears when functioning with the fine-grained data. It is noticed that the many studies joined many sources of the information, which enables 360 points of degrees the study of the object. This might involve the big structured data sets of unstructured data such as web data, tweets, text, newsfeed social media, and social bots. Diversification of the information needs a hybrid of two methods or more than two methods to analyze and process the information. The characteristic example of using machine methods is determining the formless data and involving them as inputs into models and statistics.

As per Haleem, Javaid, Khan, and Vaishya (2020), data sources are compulsory in relation to enhancing predictive performance. Considering the complexity and variability of the big data, studies in the area have the tendency to compare the output of the many algorithms or joined a few algorithms to optimize the predictive power. By choosing the algorithm which is having the best performance is not straightforward. It has been determined that no particular can be fit cases. It is being suggested that a proper model should be selected that offers the best balance between variance and bias (Agbehadji, Awuzie, Ngowi, & Millham, 2020).

3.1 Deep learning applications for COVID-19

The definition of which technique of detailed learning is the use of network switch is neural and one or more than two layers. Network, a neural layer, is a typical "layer" made of the parametric transformation of a nonlinear input. Moreover, it has been determined the deep learning is considered an

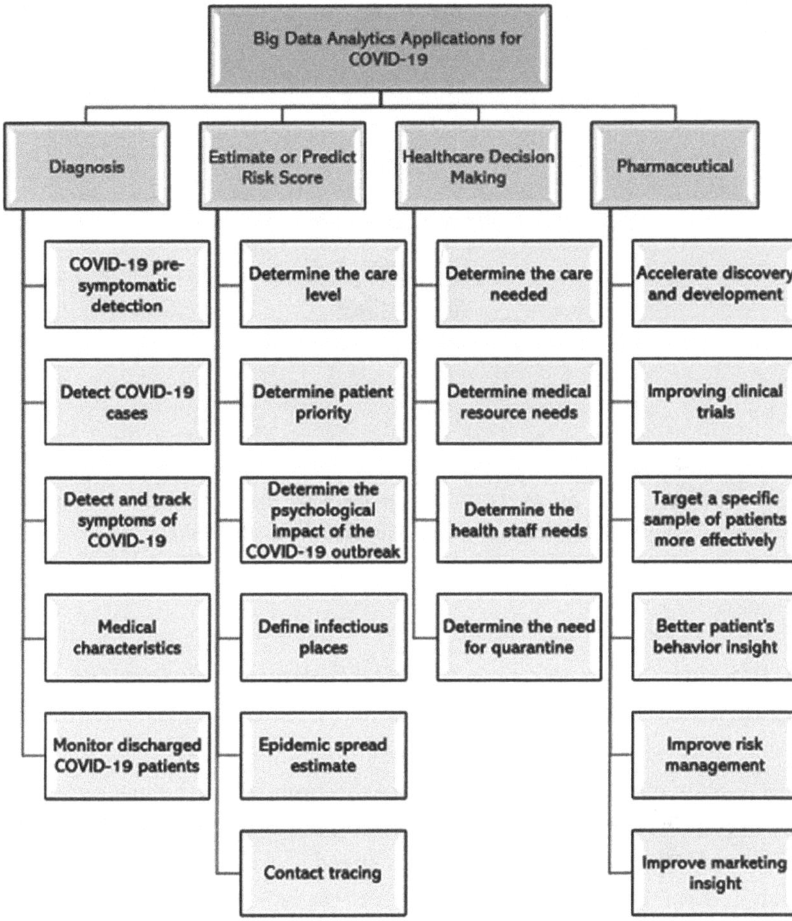

Fig. 4 Big data analytics in COVID-19.

important application in COVID-19, including misinformation detection preparation mining and public sentiment analysis. Computer vision is also considered an important application to the domain in relation to deep learning. The solicitation of deep learning can help in fighting the promising coronavirus, but it is considered quite significant to be acquainted in relation to the disadvantages of deep learning. The main issues are related to data privacy, interpretability, learning from the limited labeled datasets and generalization matrix. It is only analyzed and is quite crucial in interpreting the output of deep learning models. See Fig. 5.

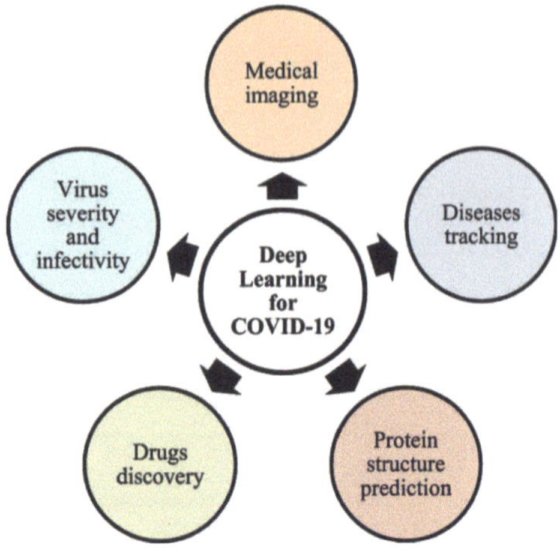

Fig. 5 Deep learning as a tool in COVID-19. *(Data from Asraf, A., Islam, M. Z., Haque, M. R., & Islam, M. M. (2020). Deep learning applications to combat novel coronavirus (COVID-19) pandemic.* SN Computer Science, 1*(6), 1–7.)*

3.2 Big data analytics as a tool for fighting pandemics: A systematic review of literature

The big data term has been become extensively used in the growth of data exponentially in recent years. Because of this popularity, the meaning of this is diverse. It can also understand as the volume of data which is huge with the diversity of topics and with the highest speed and it will also in the variety which is coming from many different sources like sensors, social networks and the cell phones and many other devices which will need the specific tools and techniques for their process.

As shown in Fig. 6 big data science study has many different aspects of big data that involve data storage capacity and analysis of the speed. Show the study of the different areas it can be employed, such as social media, commercial activity and health data. This study has the common thing, which is needed for the analysis of the data symbol analysis will not have the configuration or the capacity which is the same that the big data analytics will provide to the users (Sheng, Amankwah-Amoah, Khan, & Wang, 2020). BDA is becoming the modern practice for data analyzing also find as the analysis, which is specific for this scenario.

Fig. 6 Big data analytics in COVID-19. *(Data from Jia, Q., Guo, Y., Wang, G., & Barnes, S. J. (2020). Big data analytics in the fight against major public health incidents (including COVID-19): A conceptual framework.* International Journal of Environmental Research and Public Health, *17(17), 6161.)*

4 Review of big data analytics, artificial intelligence and nature-inspired computing models towards accurate detection of COVID-19 pandemic cases and contact tracing

Contact tracing with big data analytics: the contact drawing through Big Data Analytics is considered as a significant domain that can result in assisting the outbreak of the disease using different sources of data that includes tags used in social media, Post with metadata, list of passengers, the logo of a means of transportation that people travel with smart cards to metro and the usage of different credit cards are all considered as appreciated sources of data.

In this article, the contact tracing model has been estimated that results in the purpose of the rate of the broadcast of infection. This model can be easily identified that the potentially infected individuals before the emergence of the system. It has been determined that the most distinguished feature of humanoid social contact is that it includes the use inconsistency in the degree and number of associates in both the number of secondary cases as well as individuals that result in matching the contract tracing.

The Big Data Architecture is quite important in ensuring the Sharing, data integration and analysis within the environment of big data. The Data Construction can result in an assessment of the real-time nursing of different public health surveillance systems. When information is collected from big data through social networks, it can easily be rebuilt in determining the outbreak of diseases (Sheng, Amankwah-Amoah, Khan, & Wang, 2020). Machine learning and Artificial Intelligence are considered the important expectations in screening the population of people that can help assess the infection risk. The temperature screening has generally been detecting the symptoms and the isolated persons expected to be infected. Moreover, it has been determined that the artificial intelligence-powered smartphone app will develop tracking the spread of COVID-19 in various Geographic areas.

5 Significant applications of big data in COVID-19 pandemic

Big data technology can help store a various quantity of information about the infected people with the virus of Corona. It generally aids in the understanding of the nature of these viruses. This technology is generally used to store the data of all types of cases pretentious by the COVID-19. This info

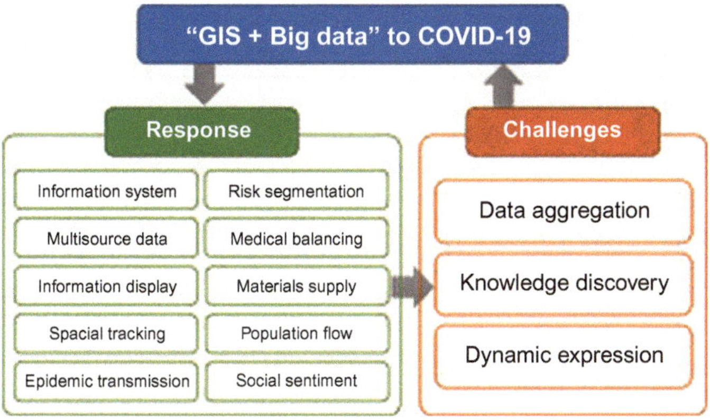

Fig. 7 Applications of big data in COVID-19. *(Data from Zhou, C., Su, F., Pei, T., Zhang, A., Du, Y., Luo, B., et al. (2020). COVID-19: Challenges to GIS with big data.* Geography and Sustainability, 1*(1), 77–87.)*

can review in identifying the cases and in helping and allocating the resources for the protection of the health of the public. See Fig. 7.

Several digital data modalities that include the location of the patient, patient-reported travel, proximity and current systems can generally be digitized and may be used in the generation of actionable insights at the demography and community level. Moreover, it has been determined that big data can help provide various amounts of evidence to health workers, epidemiologists, and scientists that can make informed decisions about fighting the virus. This data can mainly be used to track the virus on a continuous Global basis and help create innovation in the medical field. It can also help forecast the impact of COVID-19 in the whole population and particular area. Big data can also help provide opportunities and possible sources for the people that can help in having this stressful situation (Shorten, Khoshgoftaar, & Furht, 2021). This technology can help provide the data that can analyze disease transmission, monitoring of the health, movement, and prevention system. Letters of the big data will be acting as the medium for searching, tracking, and controlling the COVID-19 in the pandemic. It will take for the manufacturing and the development of the vaccine with having the knowledge. The model data helps to understand and offer the edge over the many other processes like corresponding homology models: product bi function and fold task service system for a specific target, dimple also is downloaded from the protein data bank, find the symptoms, and predict the COVID-19 connected with it. Big data analysis can better understand how the transmission could have happened and provide up-to-date updates and insights to both the stakeholders and the general public. With the help of data collection and analysis, together with lockdown, China suppressed the COVID-19 and implemented it with the help of artificial intelligence towards the spreading at a low rate.

The journal of Chinese medical association the article on the outbreak of COVID-19 was written by Wu, explain the first arrival of this virus in late 2019 December, the emergence of this virus is Wuhan, China, it out broke in some cities of China and then spread globally including the countries of Japan, United States, Philippines, Vietnam, India and many more. According to the ministry of health, infection is recognized as coronavirus disease or COVID-19 by WHO. This infection is a communicable disease under category 5. This disease is potentially zoonotic with a mortality rate to moderate. Its transmission may be spread through the droplets any kind of contact or if there is any kind regarding the control of stringent infection, or if there is no personal protective equipment. Presently there is no treatment for this virus all though the investigation is continuous (Tang, Schmitz, Persing, & Stratton, 2020).

With the continuous spreading of the infection worldwide, rapid tools of diagnosing, always handy repurposable drugs and instant containment measures Stopping the COVID-19 pandemic: a review on the advances of diagnosis, treatment, and control measures are of paramount importance. MD. Hasanul Banna Siam, in the journal of pathogens, explains that various kinds of vaccines, for example, RNA, DNA, and LENTIVIRAL, inactivated and viral vector is in trials in clinics. The very rapid deployable and innovative vaccine delivery systems have also been developing. So many countries have started going easy on the measures of lockdown, but the second wave of infection will need very strict as well as rational control policies so that the fatalities could be minimized (Siam, Banna, Nishat, Ahmed, & Hossain, 2020).

In the journal of clinical microbiology, the article on laboratory diagnosis of COVID-19: current issues and challenges elaborates that the outbreak of COVID-19 has had a very wide impact on clinical microbiology laboratories in the past several months. The current issues and challenges regarding the laboratory diagnosis of the infections caused by the very dangerous acute respiratory syndrome coronavirus 2 (SARS-COV-2) are widely covered. It has been analyzed that during the pre-analytical stage, the collection of the proper specimen of the respiratory tract at the right time and from the right anatomic site is essential for a fast and apt diagnosis of COVID-19.

In this article, a systematic review and meta-analysis have been done to investigate the optimum distance so that person-to-person transfer of the virus could be avoided and extensive uses of face masks and eye protection can largely reduce the transmission of this virus. This data has been searched through various data sources from the third of May 2020 inception with your language restriction for the studies and feasibility resource use and equity. The findings showed that there would be a lower transmission of this virus because physical distancing of at least 1 m or more will be maintained compared to a lower distance than that. The use of face masks resulted in a significant reduction in the risk of infection, the protection of the eye is also related to less infection (Chu et al., 2020).

Ahmad Elragal wrote on big data analytics in a literature review paper that huge amounts of data are available to decision-makers on hand in this era of information. Big data is referred to as data sets. These data sets are used to have very high amounts of velocity and variation, which makes them hard for handling using the same old-style tools and techniques. Because of such quick growth in data, solutions should be studied and should also be provided in audio for handling extracting value as well as knowledge from datasets. Further, Big data analytics is a complicated process through which examination of big data is done to the revelation of information, for example, patterns that are not very clear, the trends regarding the market, the preferences of customers, which will be helpful to organizations for taking precise and apt decisions (Tsai, Lai, Chao, & Vasilakos, 2015).

With new age of big data which is impossible to be handled by the traditional data analytics due to huge quantities of data (Big data analytics: a survey by Chun-Wei Tsai, Chin-Feng Lai, Han-Chieh Chao, and Athanasius V. Vasilakos). The question now is how a great performance platform could be developed so that big data could be analyzed efficiently. An appropriate algorithm of mining could be designed to extract useful information from big data. On a wider scale, the technologies of data analytics and the techniques are helpful to organizations by providing a way in which analyzing data sets and gathering new and significant information could be done. The queries regarding intelligence can help answer the basic questions in relation to operations and the performance of the business (Elgendy & Elragal, 2014).

In this paper written by Bala M. Balachandran, the benefits and challenges included in the deployment of big data analytics are outlined. It is argued that the storage and computing necessities of big data analytics are supported by cloud computing. The combination of these very important technologies will help in the enhancement process regarding big data mining improve the decision-making processes of the business. Organizations use big data analytics systems and software to make decisions that are data-driven so that there is improvement in the outcomes related to the business. The advantages include better marketing, greater opportunities for revenue generation, personalization of customers, and improved efficiency of operations. And these all benefits can be helpful in creating competitive advantages over the rivals (Balachandran & Prasad, 2017).

Machine learning is a method of artificial intelligence for discovering knowledge regarding making effective and intelligent decisions. Big data has huge effects on the value creation and discoveries related to science. Wang and Alexander (2016) introduce the methods in machine learning and the important technologies in big data and some kinds of applications inventor of machine learning. There has been a paradigm shift and the strategies of the firms globally, which is brought by social media. A resultant of it is the collection and message from so many varieties of social media platforms that have necessitated using this data for business intelligence purposes. This study has filled the gap in knowledge through the examination of the rules and implications regarding big data analytics business intelligence (Ram, Zhang, & Koronios, 2016).

The pandemic of COVID-19 has been labeled like the Dark swan event since it has affected a huge impact on almost all aspects of human life. The findings and the propositions in the study suggest that COVID-19 will become the substance of various long and short-term changes in policy and will also require researchers' hypothetical and pragmatic attention as published in the journal of business research (Verma & Gustafsson, 2020).

The noble coronavirus has spread to 210 countries worldwide which was declared a pandemic. It had a significant impact on the health systems, the economy, education, and social facets of present-day society. Since the transmission rate increased, many collaborative approaches in the stakeholders have developed innovative means of screening, detecting, and diagnosing coronavirus cases in humans. The importance of the computing models related to the 4th industrial rebellion technologies in achieving the wanted feat has been emphasized. However, there has been a wide gap in the two accuracies of the detection and the protection of the coronavirus cases. The tracing of the contacts of the person who is infected is also difficult. The following paper represents a review of the computing models that can be adopted to enhance performance and predict global pandemic cases that are COVID-19. Tata official intelligence focuses more on big data, which is nature-inspired computing included in the present pandemic. The review has recommended that these non-natural intelligence models be used to detect the geysers of COVID-19. In the same way that forms of big data can also be applied for contact tracing. Nature-inspired computing models have illustrated amazing performances in the feature selection of the issues which are still open to exploration (Agbehadji et al., 2020).

Although the management scholars do distinguish the value of using the big data to comprehend, foresee and give response to upcoming events, there is a very inadequate overview of how so many different analytical techniques could be harnessed to help the guiding scholars in the researches and the global challenges raised in the pandemic related to COVID-19. Indeed the advancements, as well as the proliferation of so many technologies, have culminated. Various insights have been divided into the methods in the diagnostic, the predictor and the prescriptive analytics. These methods could be liberated to study "Black swan" events like the COVID-19 worldwide crisis and its various insinuations to the managers and policymakers aftermath (Sheng, Amankwah-Amoah, Khan, & Wang, 2020).

6 Research problem

The research problem of this project is dealing with data analysis and data mining during COVID-19 through machine learning in the addition of deep learning of procedure which is faced situations of COVID-19. In the current scenario of the epidemic, the people facing problems to analyze the existing solutions or strategies of COVID-19 impacts to minimize this impact can utilize new technologies like big data analysis with the help of machine education. Through the help of new technologies, data analysis and data mining of the pandemic is easy to identify and provides better outputs to solving problems of coronavirus disease.

7 Research questions

- What is the part of Machine Learning while dealing with the situation of COVID-19?
- How can a study of COVID-19 be leveraged by using Big Data Analytics and its method?

8 Objectives

The main objectives of this project are:

- **To explain the role of machine learning**

  The role of machine learning techniques to tackle COVID19 crisis: A systematic review

  They generally greatly alter the symptoms of a coronavirus that resulted in fluctuating from mild flu to severe respiratory suffering. The adoption of machine learning techniques has resulted in employing different skates that generally range from analyzing the disease spread. The various data types, including radiological images, drug databases and collection of the data from public health agencies, have generally been used for the prediction. It is only believed that machine learning can help predict all the three risks, such as infection risk, severity risk, and outcome risk. Machine learning founded CT analysis is also considered as the showing tool in the occurrence of COVID-19. It is required that the algorithms are required to be skilled in the full range of disease that includes asymptomatic. The occurrence of the COVID-19 has resulted in offering various exciting opportunities for the applied artificial intelligence research, as shown in Fig. 8 (Syeda et al., 2021).

- **To explain the various methods of data collection**

  Entities can use big data analytics and similar software to make business decisions, which can help improve the outcomes related to those decisions. The advantages of them involve many new opportunities related to the generation of revenue, the personalization of customer experience, new and effective techniques of marketing, improvement in operational efficiency. With the use of an effective strategy, the benefits could be helpful in the creation of a competitive advantage over its rivals. The working of big data analytics could be explained in the following way: data analyst, the scientists of data, arithmeticians besides other professionals collect, then course, clean as well as examine the increasing capacities of the deal data which is structures and also many other forms of data which are not put into use by the analytics programs and the conventional business intelligence. The process of preparation of data is explained in the following four steps:

  - Data collection is done by the professionals of data from various sources. Most times, it is like a mixture of data which is semi-structured and unstructured. Though every organization has different streams of data, some very common sources involve web server blogs, mobile

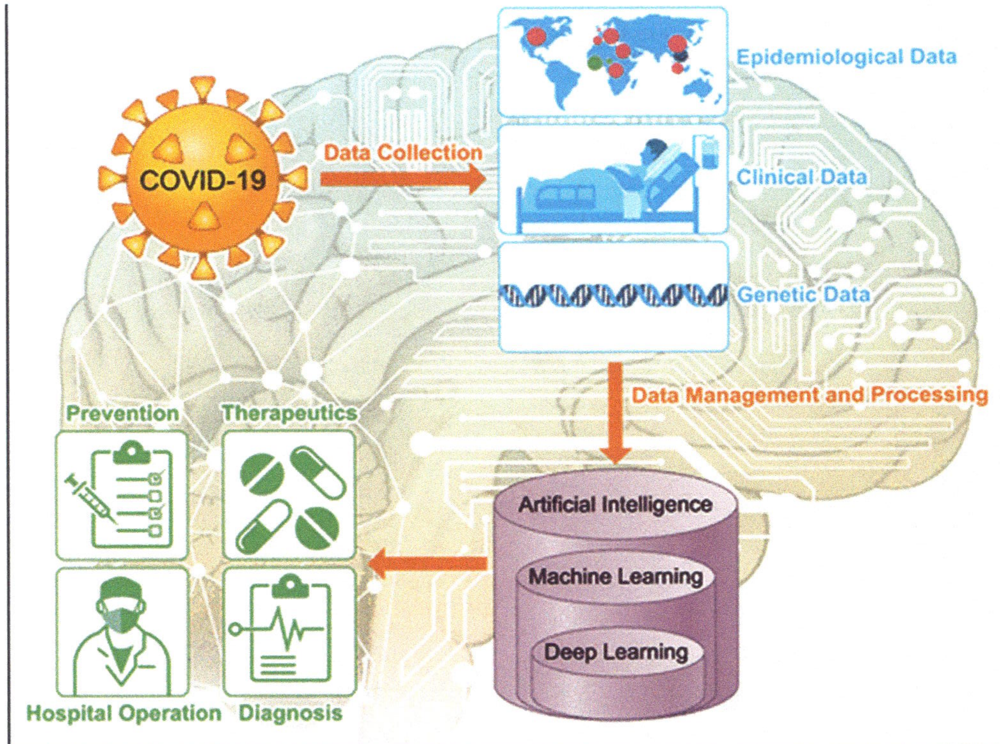

Fig. 8 Role of machine learning techniques to tackle COVID19 crisis. *(Data from Alimadadi, A., Aryal, S., Manandhar, I., Munroe, P. B., Joe, B., & Cheng, X. (2020). Artificial intelligence and machine learning to fight COVID-19.)*

applications, telephone records, the mechanism data taken by the devices linked to the internet of things, various cloud applications, and so on.

- Processing of data. Once the collection and storage of data are done in the data warehouse, its organization must be done by the professional of data. The partition, as well as the configuration of data, must be done properly for analytical queries. Proper processing of data contributes to higher performance from the analytical queries.
- In the third step, the cleaning of the data is done for better quality. Various tools of scripting or the software of the enterprise are used by the data professionals for scrubbing data. They try to find any errors or inconsistencies present, such as the duplication of data, any mistakes in formatting, organization or tidying up the data.

• **To explain different algorithms utilize in machine learning**

Some very popular methods of machine learning are as follows:

- Supervised learning: the uses of labeled examples train these algorithms. For example, for an input name, the output that is desired is already known—an example of an equipment piece that might have points of data labeled. Moreover, F means unsuccessful and R means runs. The learning algorithm receives a set of inputs along with the correct corresponding output, and learning in this algorithm is done by creating a comparison of the actual output with the correct output for finding errors. The modification of the model is then done accordingly. By using approaches like sorting, forecasting, and gradient boosting unsupervised wisdom, the designs are protected by the values regarding the label on the unlabeled data, which is extra. Supervised learning is most commonly used in applications where the prediction of historical data is likely to predict upcoming events. For instance, anticipation could be done when dealings in the credit card could be fraudulent or when the customer of insurance has chances of filing a right. See Fig. 9.
- Unsupervised learning: it is cast-off as in contradiction of the data with no historical labels. No "right answer" is expressed by the system. It is up to the algorithm regarding what is to be shown. The main goal is the exploration of the data and searching for some kind of structure within. This kind of unsupervised learning is fine suited to transactional kinds of data. Given examples, identifying a segment of customers with very akin attributes can be treated in a similar way when it comes to marketing campaigns. See Fig. 10.

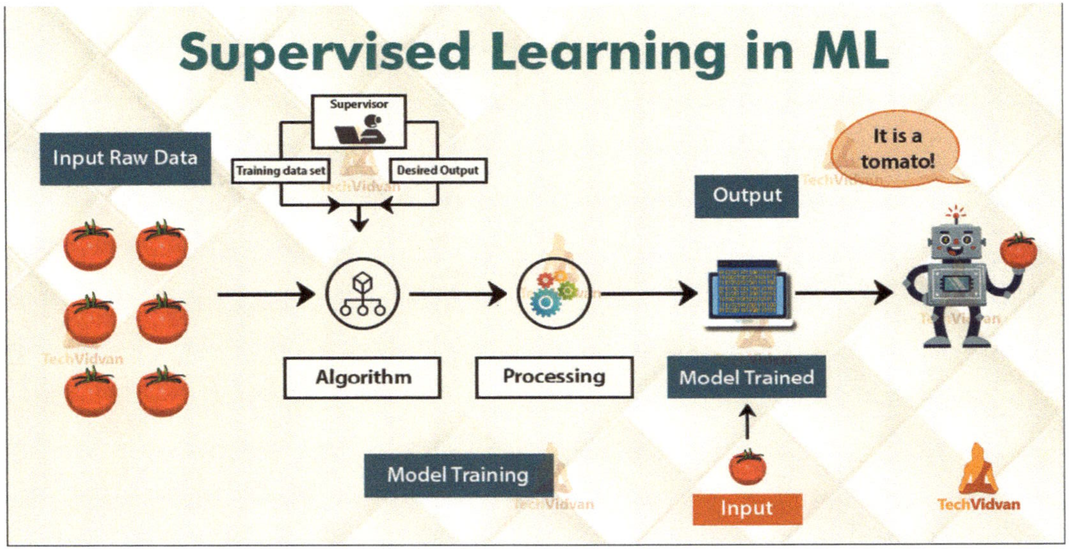

Fig. 9 Supervised learning. *(https://www.linkedin.com/pulse/brief-review-artificial-intelligence-its-subfields-dwaipayan-saha/.)*

– Reinforcement learning: this is furthermost used for gaming, steering as well as robotics. Through the methods of trials and errors, this algorithm can discover what kinds of actions will help yield the biggest recompenses. There are three main components regarding this kind of learning which are: the agent (the choice maker), the atmosphere (everything is interacted by the agent) and the actions (everything in which the agent can do). The main purpose for the agent is to decide which actions will help maximize the reward, which was expected over a specific period. See Fig. 11.
– Semi-supervised learning: its use is similar to that of supervised learning, but the only difference is that both labeled and unlabeled data are used for the training, mostly some quantity of data labeled in combination with a huge amount of data quantity of data which is unlabeled. The usage of this learning is in various methods like classification, prediction as well as regression. When the cost related to the labeling is very high such that a fully labeled process of training could be allowed, in such situations, semi-supervised learning is very effective and useful. See Fig. 12.

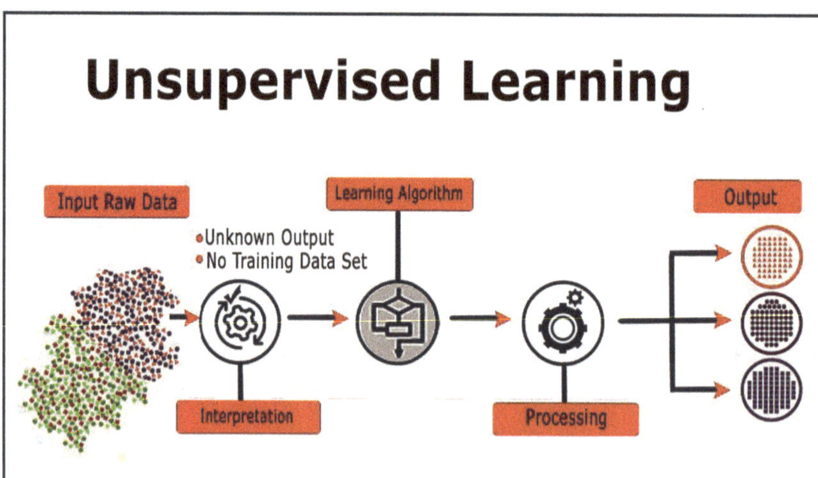

Fig. 10 Unsupervised learning. *(https://towardsdatascience.com/into-to-pca-k-means-clustering-unsupervised-learning-algorithms-5cc5acea274d.)*

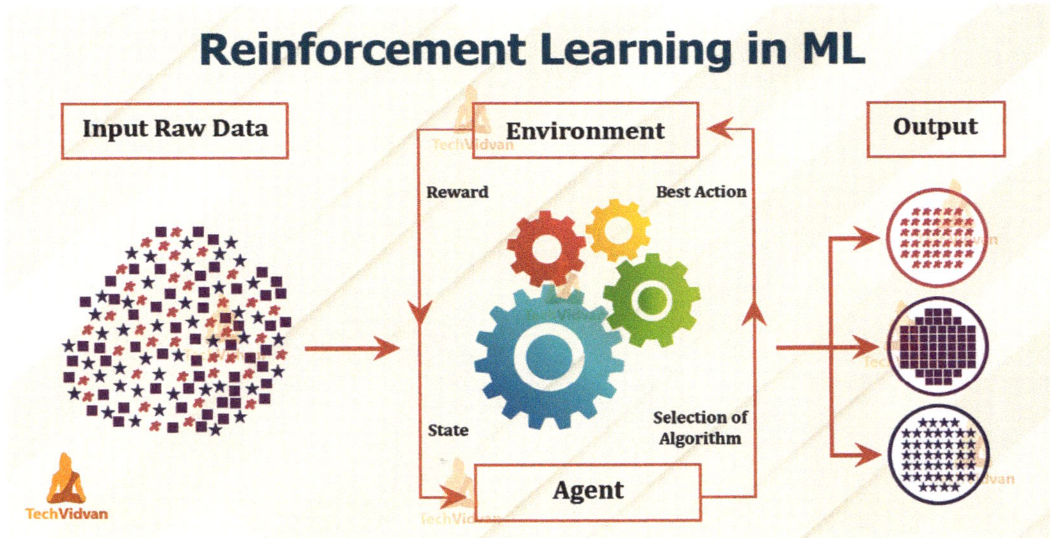

Fig. 11 Reinforcement learning in ML. *(https://techvidvan.com/tutorials/reinforcement-learning/.)*

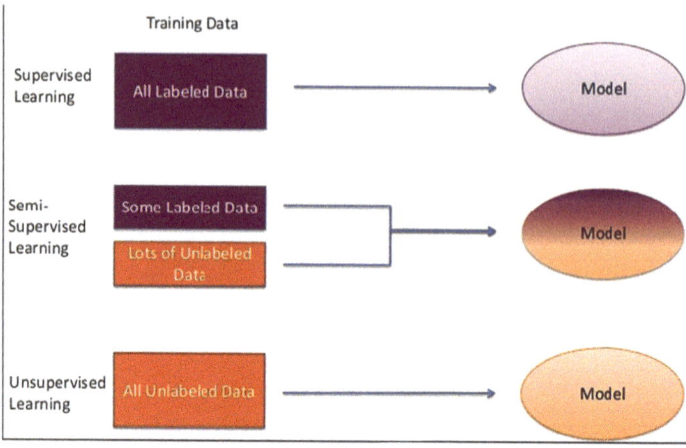

Fig. 12 Semi-supervised learning. *(https://www.analyticsvidhya.com/blog/2017/09/pseudo-labelling-semi-supervised-learning-technique/.)*

- **To explain the utilization of BDA and machine learning in the situation of COVID-19**

  According to the author (Jia et al., 2020) design and development of new generation architecture and technology can be helpful for BDA or Big Data Analytics to get maximum velocity and examine and discover various variations and high volume of data for extracting values from them. The increasing popularity of information technologies such as big data technology, social networks, cloud computing or IoT, or the Internet of Things has been used widely in healthcare, supply chains, logistics, and other industries. BDA is helpful and convenient in decreasing fear by revealing the epidemic situation in real-time and clarifying rumors regarding the COVID-19. Model of BDA or analysis of big data can be used to estimate the impact of this situation on social development, economic, political development and assist the government in making suitable changes.
- **To explain domain adaptation and big data**

  As per the words (Jia et al., 2020) of three mechanisms that can be used for controlling and preventing COVID-19 situation by big data application are prevention of situation, response to situation and recovery from the situation. With the use of visual analysis technology, the relationships between massive datasets can be uncovered. It enables the investigators to attain visual cognition that is more intuitive and sufficient support in the decision-making process. Visualization is achieved with the help of GIS or Geographical Information Systems. The government uses sources of bid data for conducting pictorial analysis of monitoring the widespread situation, hospital enterprises, medical resources, and screening for making decisions. Government can improve mechanisms used by them for an epidemic response that are based on analytics regarding big data. Big data technology is used to identify disasters, supporting the decision, communication and coordination, and technical support. For identification of disaster, predictive scrutiny of dynamic models of infectious disease is combined with data to predict how critical a situation is and provide support. Systems of geographic situation and examination of Graph database are used for tracking persons who are infected and who are their contacts. Government can establish a mechanism for repairing an epidemic based on analysis of big data. Big data is used for eliminating fear, audit assessment, recovery and adjustment of policy.

9 Methodology

In this research, Experimental Research methodology is used to find the best algorithm. An experiment is a kind of research method in which there will be manipulation of one or more independent variables and missing their impact on one or more dependent variables. An experimental design is a process of the development of sets of procedures for testing the hypothesis. Experimental research methodology is being conducted with the help of a scientific approach by utilizing two sets of variables. The first set will play a role of a constant that has been used for measuring the differences of the second set. This research methodology will help establish a noticeable effect and cause. This type of research methodology will help collect information from present and potential customers by using the sampling method. With the help of various kinds of surveys, questionnaires and the outcomes of those will be depicted in the form of numbers. See Fig. 13.

The data collection methods used for current research involve no data collection and the algorithms discussed have been sourced from various existing literature.

9.1 Techniques

The data and other relevant information are gathered through secondary search. It is a kind of research method which includes the utilization of already present data. The data present has been summarized and collected to enhance the overall effectiveness of the research. This search involved research material that is being published in a research report and similar kinds

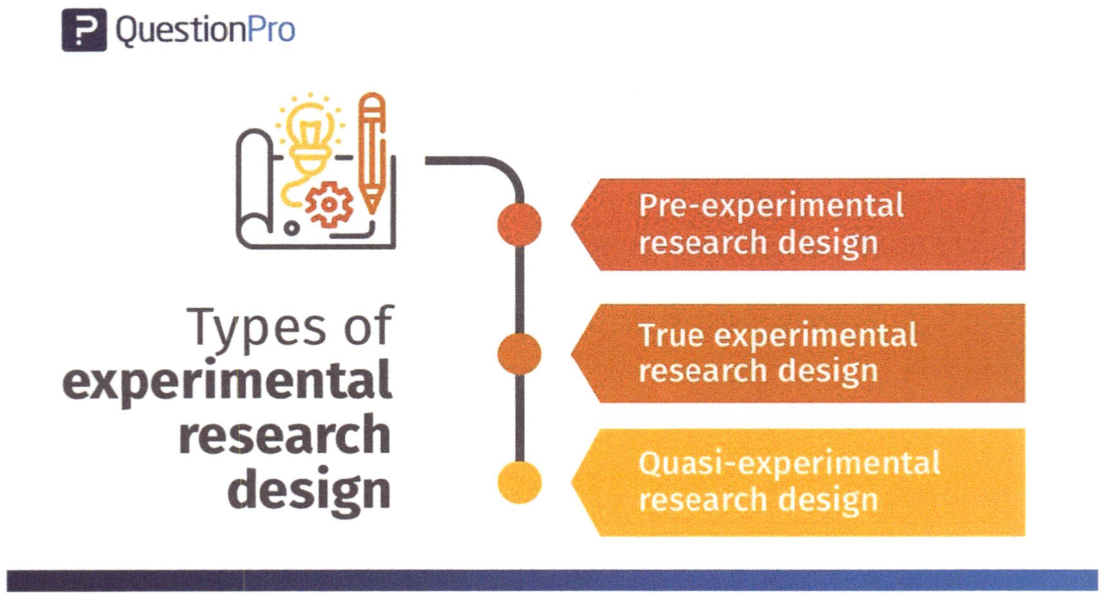

Fig. 13 Experimental design. *(https://dewezanake.vercel.app/index/view-17-one-shot-case-study-research-design-example/.)*

of documents. It is more cost-effective than primary research as it makes utilization of already present data.

10 Algorithm

The algorithm that has been used in this research is Naive Bayes. It is a kind of simple and supervised machine learning algorithm that utilizes the Bayes theorem that involves stronger and independent assumptions among the features for procuring results. This algorithm assumes that every input variable is an independent variable. It is a kind of naive assumption for making results about real-world data. See Fig. 14.

This algorithm is a kind of classification algorithm based on the Bayes theorem. It is not a sole algorithm but involves the family of algorithms where all algorithms share a common principle: each pair of features is classified independently. Bayes theorem states that the probability of occurrence of an event is given by the probability of another event that exists already. See Figs. 15 and 16 for formulas.

Naïve Bayes is generally a classification algorithm based on the Naïve Bayesian Algorithm and strong Independence assumptions. This type of algorithm simplifies learning by undertaking features that are independent of provided classes. Naive Bayes is a type of Bayesian decision theory subset. The name naive demonstrates the formulation undertakes some naïve assumptions. Naïve Bayes is among the simpler probabilistic classifier. This displays amazingly well in several Real-world applications. The Naïve Bayes classifier communicates to the Bayesian network that can be demonstrated with the below equation.

Classification is generally a fundamental issue within data mining and machine learning. In classification, the major learning algorithm goal is to develop a classifier given a training set of examples with class labels. Such algorithms are utilized in diverse real-life scenarios such as spam filtration, sentiment analysis, text classification, and recommendation system. The advantage of the Naïve Bayes algorithm is that it is a relatively easier algorithm to build and understand. It is also faster to predict classes compared to other classification algorithms, and it can be easily and with the use of a small data set. In general terms, this classifier assumes that the existence of a certain feature in class concerns the existence of another feature. The Naïve Bayes model is easy to use and build for large datasets and its simplicity. The Naïve Bayes is called for outperforming the highly sophisticated classification methods. During the Independence assumption holds,

Naive Bayes

In machine learning, naive Bayes classifiers are a family of simple "probabilistic classifiers" based on applying Bayes' theorem with strong (naive) independence assumptions between the features.

$$P(A|B) = \frac{P(B|A)\ P(A)}{P(B)}$$

using Bayesian probability terminology, the above equation can be written as

$$\text{Posterior} = \frac{\text{prior x likelihood}}{\text{evidence}}$$

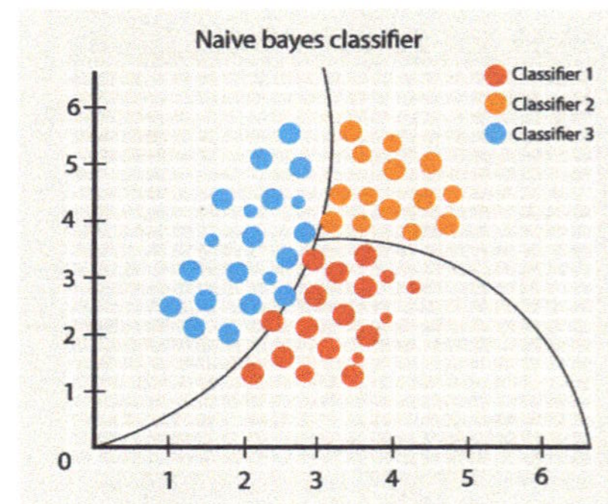

Fig. 14 Naïve Bayes. *(https://coderzcolumn.com/tutorials/machine-learning/scikit-learn-sklearn-naive-bayes.)*

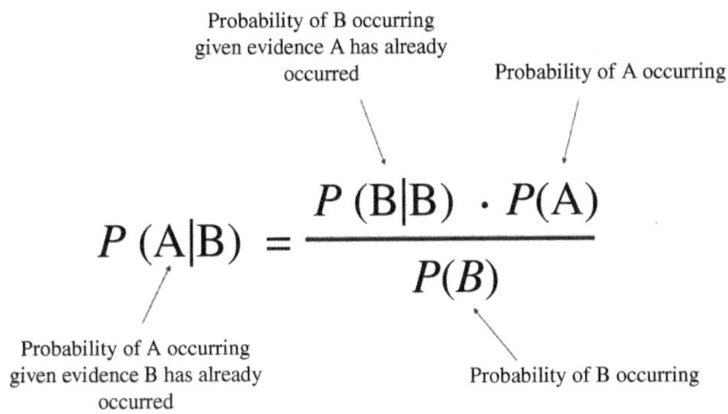

Fig. 15 The formula used.

$$Pr(c, x_1, \ldots, x_m) = Pr(c) \cdot \prod_{i=1}^{m} Pr(x_i \mid c)$$

Fig. 16 Naïve Bayes.

this classifier performs a suitable comparison to the other models such as Logistic regression and one further requires fewer data. This algorithm performs well in the situation of categorical input variables compared to the other type of numerical variables. Other applications of such Naïve Bayes algorithm include multiclass prediction and real-time prediction. Moreover, the Naïve Bayes algorithm is not always consistent and dependent on the quality of good datasets as the limitation. Please see Fig. 17 for the input and output of Naïve Bayes analysis and Fig. 18 for showing different results and analyses for COVID-19.

11 Conclusion

11.1 Big data

Big data signals an emerging emphasis on the rights of the data-centric approach to research that has to integrate, mobilize, visualize, and

Formula used

```
In [3]: from sklearn.preprocessing import LabelEncoder
        le = LabelEncoder()
        X[:,0] = le.fit_transform(X[:,0])

In [4]: from sklearn.model_selection import train_test_split
        X_train, X_test, y_train, y_test = train_test_split(X, y, test_size = 0.20, random_state = 0)

In [5]: from sklearn.preprocessing import StandardScaler
        sc = StandardScaler()
        X_train = sc.fit_transform(X_train)
        X_test = sc.transform(X_test)

In [6]: from sklearn.naive_bayes import GaussianNB
        classifier = GaussianNB()
        classifier.fit(X_train, y_train)

Out[6]: GaussianNB()

In [7]: y_pred  =  classifier.predict(X_test)

In [8]: y_pred

Out[8]: array([0, 0, 0, 0, 0, 0, 0, 1, 0, 0, 0, 0, 0, 0, 0, 1, 0, 0, 1, 0, 0, 1,
               0, 1, 0, 1, 0, 0, 0, 0, 0, 0, 1, 0, 0, 0, 0, 0, 0, 1, 0, 0, 0, 0,
               1, 0, 0, 1, 0, 1, 1, 0, 0, 1, 1, 0, 0, 0, 1, 0, 0, 1, 0, 0, 0, 1,
               0, 0, 0, 0, 1, 0, 0, 0, 0, 0, 0, 0, 1, 1], dtype=int64)

In [9]: y_test

Out[9]: array([0, 0, 0, 0, 0, 0, 0, 1, 0, 0, 0, 0, 0, 0, 0, 0, 0, 0, 1, 0, 0, 1,
               0, 1, 0, 1, 0, 0, 0, 0, 0, 1, 1, 0, 0, 0, 0, 0, 0, 1, 0, 0, 0, 0,
               1, 0, 0, 1, 0, 1, 1, 0, 0, 0, 1, 1, 0, 0, 1, 0, 0, 1, 0, 1, 0, 1,
               0, 0, 0, 0, 1, 0, 0, 1, 0, 0, 0, 0, 1, 1], dtype=int64)
```

(A)

Fig. 17 cont'd

```
In [3]: from sklearn.preprocessing import LabelEncoder
        le = LabelEncoder()
        X[:,0] = le.fit_transform(X[:,0])

In [4]: from sklearn.model_selection import train_test_split
        X_train, X_test, y_train, y_test = train_test_split(X, y, test_size = 0.20, random_state = 0)

In [5]: from sklearn.preprocessing import StandardScaler
        sc = StandardScaler()
        X_train = sc.fit_transform(X_train)
        X_test = sc.transform(X_test)

In [6]: from sklearn.naive_bayes import GaussianNB
        classifier = GaussianNB()
        classifier.fit(X_train, y_train)

Out[6]: GaussianNB()

In [7]: y_pred  =  classifier.predict(X_test)

In [8]: y_pred

Out[8]: array([0, 0, 0, 0, 0, 0, 0, 1, 0, 0, 0, 0, 0, 0, 0, 1, 0, 0, 1, 0, 0, 1,
               0, 1, 0, 1, 0, 0, 0, 0, 0, 0, 1, 0, 0, 0, 0, 0, 0, 1, 0, 0, 0, 0,
               1, 0, 0, 1, 0, 1, 1, 0, 0, 1, 1, 0, 0, 0, 1, 0, 0, 1, 0, 0, 0, 1,
               0, 0, 0, 0, 1, 0, 0, 0, 0, 0, 0, 0, 1, 1], dtype=int64)

In [9]: y_test

Out[9]: array([0, 0, 0, 0, 0, 0, 0, 1, 0, 0, 0, 0, 0, 0, 0, 0, 0, 0, 1, 0, 0, 1,
               0, 1, 0, 1, 0, 0, 0, 0, 0, 1, 1, 0, 0, 0, 0, 0, 0, 1, 0, 0, 0, 0,
               1, 0, 0, 1, 0, 1, 1, 0, 0, 0, 1, 1, 0, 0, 1, 0, 0, 1, 0, 1, 0, 1,
               0, 0, 0, 0, 1, 0, 0, 1, 0, 0, 0, 0, 1, 1], dtype=int64)
```

Output

```
In [13]: from sklearn.metrics import confusion_matrix,accuracy_score
         print("The Confusion Metrix : ")
         print(confusion_matrix(y_test, y_pred))
         print("The Accuracy Score is : ",accuracy_score(y_test,y_pred))

         The Confusion Metrix :
         [[56  2]
          [ 4 18]]
         The Accuracy Score is :  0.925

In [ ]:
```

Analysis

(C)

Fig. 17 Inputs and command-line outputs of running Naïve Bias (A and B).

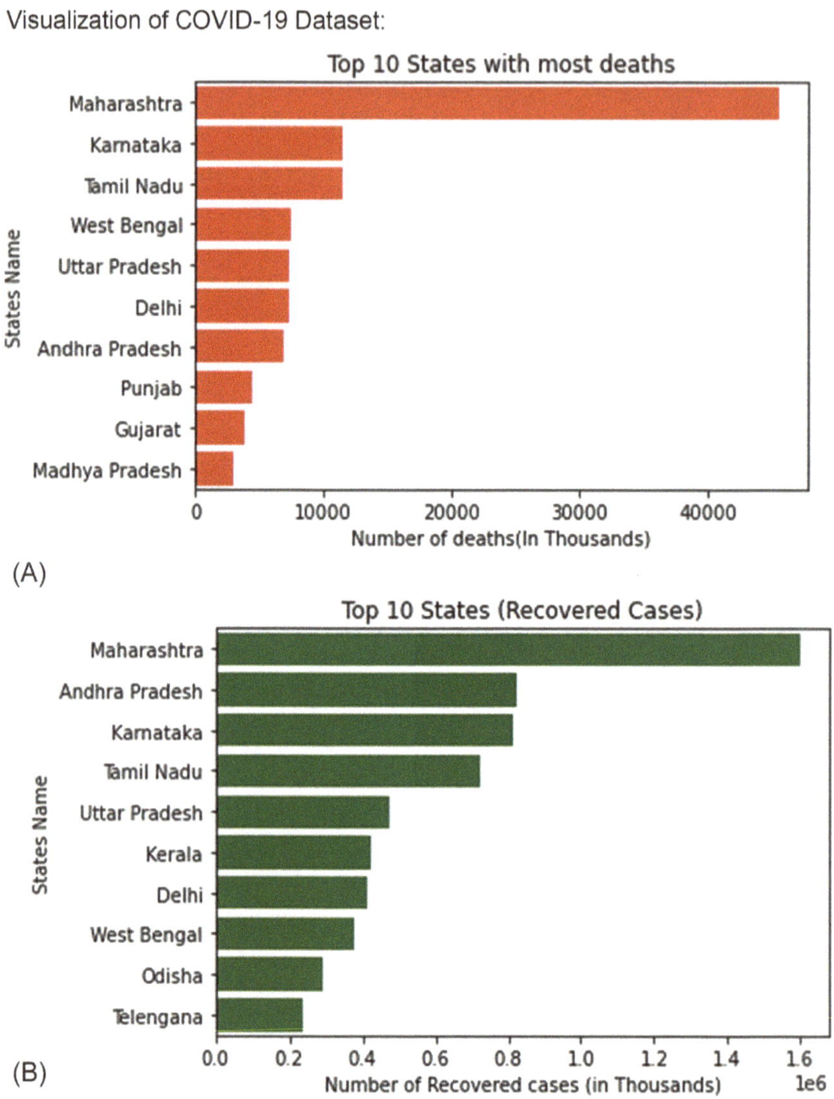
Visualization of COVID-19 Dataset:
Top 10 States with most deaths
Maharashtra
Karnataka
Tamil Nadu
West Bengal
Uttar Pradesh
Delhi
Andhra Pradesh
Punjab
Gujarat
Madhya Pradesh
States Name
0
10000
20000
30000
40000
Number of deaths(In Thousands)
(A)
Top 10 States (Recovered Cases)
Maharashtra
Andhra Pradesh
Karnataka
Tamil Nadu
Uttar Pradesh
Kerala
Delhi
West Bengal
Odisha
Telengana
States Name
0.0
0.2
0.4
0.6
0.8
1.0
1.2
1.4
1.6
Number of Recovered cases (in Thousands)
1e6
(B)

Fig. 18 Cont'd

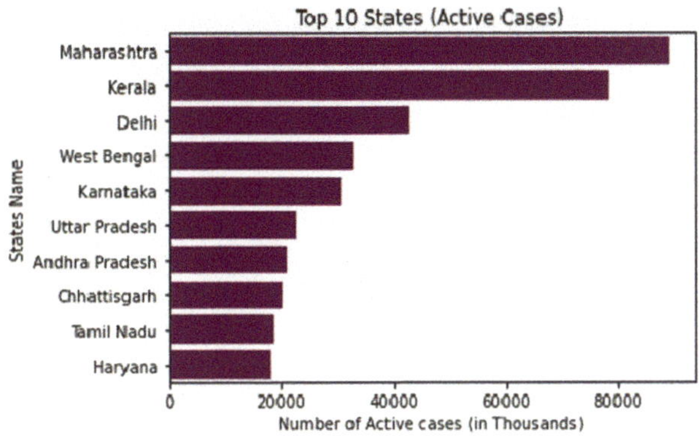

(C)

State	Recovered	Deaths	Confirmed	Active
Maharashtra	1597255	45560	1731833	89018
Karnataka	811581	11453	853796	30762
Andhra Pradesh	820234	6828	847977	20915
Tamil Nadu	720339	11415	750409	18655
Uttar Pradesh	473316	7281	503159	22562
Kerala	422410	1771	502719	78538
Delhi	410118	7228	459975	42629
West Bengal	376696	7452	416984	32836
Odisha	292477	1469	305000	11054
Telengana	235950	1393	254666	17323
Bihar	216940	1162	223917	5815
Rajasthan	198139	2019	217151	16993
Assam	203308	954	209633	5371
Chhattisgarh	183190	2507	205923	20226
Haryana	170250	1960	190323	18113

(D)

Fig. 18 Cont'd

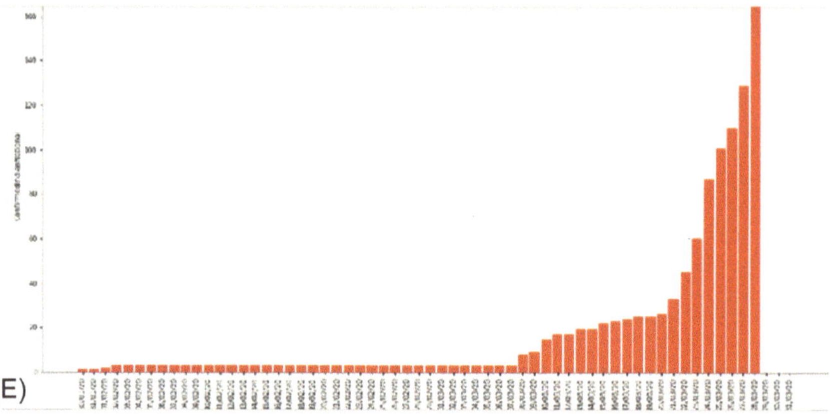

(E)

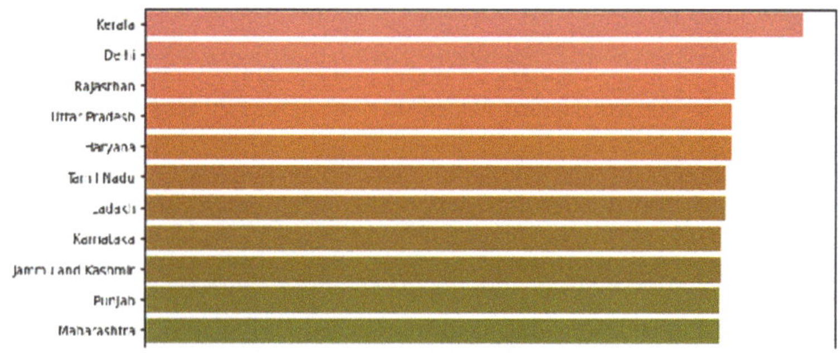

(F)

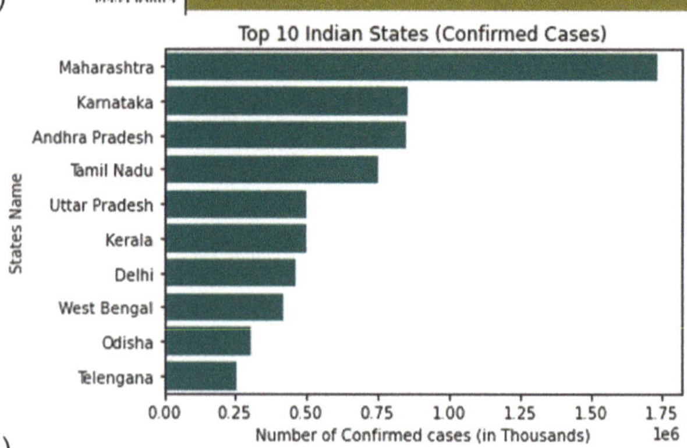

(G)

Fig. 18 Cont'd

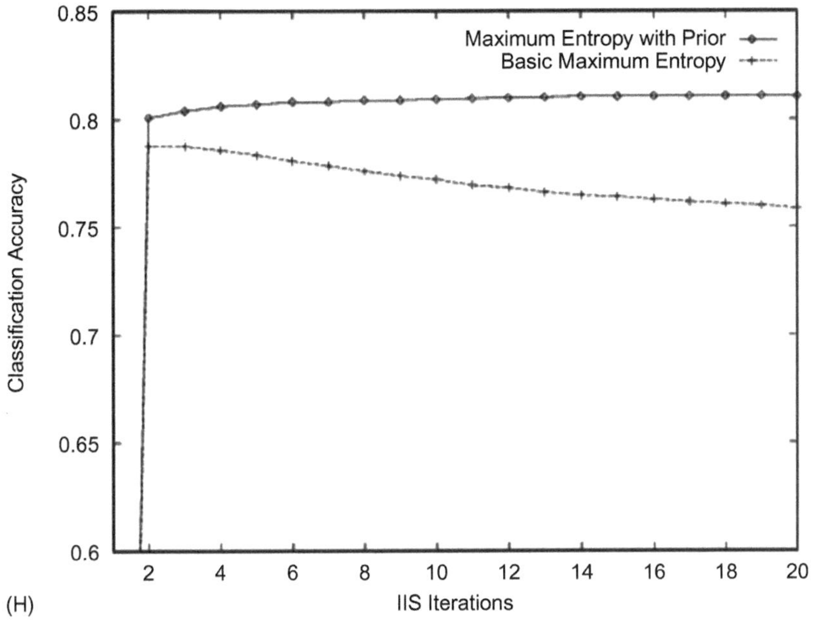

(H)

Fig. 18 Different results and analyses for COVID-19 (A–H).

disseminate the data that is viewed as contributing to the discovery. The exposure of the data centrism highlights the challenges involved in classifying, gathering and data interpretation, technologies, and the concept surrounding this process—tools such as applications and measurement instruments that produce a huge amount of data in the digital format. Data are available immediately for mentioning through the online platforms available or accessible to anybody with a broadband connection in a matter of seconds. The data access is carried through technical, ethical conceptual and legal implications and even if the access is available. The computational and mathematical tools developed to determine the big data or an opaque in their assumption and functioning lead to credibility and the scientific meaning that may difficult to access. This resulted in increasing the very related to the big data science that may be grounded upon and result in supporting the process of human making ingenuity Hostage and ultimately unintelligible intelligence.

Moreover, it has been determined that the most confronting expectation in relation to big data science has been discussed in this article: the extent to which it results in a deviation from an understanding of rationality based on individual agency and abilities. The power of particular data set results in ending the knowledge that generally lies in the extent to which it is linked to others. It results in high epistemic value to digital objects such as sequencing data on GPS location and what generally makes aggregation of data from a wide variety of sources into an effective surveillance tool. Dissemination channels and data products such as government databases, mission media and Research repositories generally read in the Interlinked and globalization distributed network that tends to function through skills and expertise. The distributed nature of making decisions is mainly involved in developing big data and analytics. It results in it impossible for any particular individual to retain oversight over the scientific significance, quality, and potential social impact of the knowledge produced.

The analysis of the big data main constitutes the ultimate instance of the cognitive system. Many groups' individuals and institutions result in ending the responsibility for the concept interpretations and outcomes of specific data use. The main key challenge related to big data government is finding the mechanism for allocating the responsibility under the acrostic from its network so that unwarranted and erroneous decisions can be taken or sanctioned. Thinking about the particular history use of the term processing can generally increase the Philosophers in avoiding historical, contextualized approaches concerning the question of evidence.

11.2 Machine learning

View detailed analysis of machine learning, and it has been analyzed that it opens up a parallel opportunity for the particular organization and enables efficiency, automation and innovation. It has been analyzed that the five pillars of Steve architected Framework based on machine learning can provide best practices in architecture for operating and building secure, reliable, efficient and cost-effective machine learning workloads in the Clouds.

Such operation automation requires teams of cross functions and areas of expertise to maintain and the major functions have full automation capabilities. For reliability, there is a robust response when there is a failure of the system by taking advantage of self-healing based on automatic scaling and keeping track of all kinds of artifacts by changing the functional system that can be automatically rebuilt.

Workloads of machine learning should be physically secured by using authorization and authentication tightly controlled with user identities, allowing users to access data and artifacts. The application will help in protecting the sensitive information of the organization and compliance requirements at every layer. Machine learning and landscape generally involve the ecosystem of tooling and growing processes and maturing.

After the coronavirus outbreak, the medical industries and scientists around the globe have worked in fighting against this pandemic and searching different production process methods and Rambha screening forecasting contact tracing and growth of the vaccine or drugs with the more reliable and accurate process. That official Intelligence and machine learning is considered a promising method that helps in improving various Health Care providers. This research has addressed Singh audition studies that generally apply different technologies to supplement the researchers in different angles and speak of challenges and trouble using different algorithms to assist the medical experts in related problems. This paper also has discussed the suggestion about conveying research on artificial intelligence-based model design policymakers and medicinal experts on this is a few and those met in the present situation while undertaking the coronavirus. Through this, it has been analyzed the use of modern technology such as artificial intelligence, and Artificial Intelligence can help refine the predictions training content using drugs and development and forecasting through extreme reliability? The power of particular data set results in ending the knowledge that generally lies in the extent to which it is linked to others. It results in high epistemic value to digital objects such as sequencing data on GPS location and what generally makes aggregation of data from a wide variety of sources into an effective surveillance tool. The application of technology in the research results in improving various algorithms and is more from this potential and advanced among the different learning algorithms. However, the current agency generally requires that there should be an improved model and the higher and performance that result in protecting and screening the coronavirus to use a different kind of diseases and analyzing its clinical and demographic info of the infected patients. Finally, it has been determined that machine learning and artificial intelligence are essential in improving the medication treatment screening and prediction and reducing the intervention of humans in the medical practice.

11.3 COVID-19

It is difficult to predict the coronavirus spreading across the world. The humanitarian health and the policies of the social economy will be adopted by different countries, which will analyze their strength and speed of recovery. Different Rock present in brief gives the country guidance as they are progressing through the different phases of crisis and also to the international community. There must be a response that is human-centered and grounded in solidarity. The International Labour values and contained the guidelines for safeguarding the decent work that is even valid in the context of the catastrophe of a pandemic. The main emphasis is on responding to the crisis that needs to ensure that human rights should be respected. And the rule of law should include the respect for principles that are basically at exertion for global labor standards. The standard guidelines for employees at work and safety may undergo public security, non-discrimination preparations of emergencies and isolation of infected workers and workers under quarantine are required to the rapid response to ease in stouter recovery from the crisis. It is required that there should be coordinated Global effort that can help in supporting the countries that can lead to and is not sufficient in fiscal space and financing the social policy in particular social protection system. It has been determined that without proper long-term structure changes, the inequalities exposed in deep-rooted crises will intensify. This will help take the immediate effect of prices and has unique opportunity to adopt a policy that thinks that achieving human-centered in Social Justice Future of work. The virus is transmitted between the people to the other person. It shows the symptoms and 2–14 days depending upon the incubation period. As per the symptoms and this sign of middle to moderate, moderate are for fatigue, fever and dry cough, and shortness of breath. Tiredness and fever also occur. Individuals with other diseases like diabetes, heart disease, and asthma are more deadly viruses and may become ill. While covering and sneezing the mouth in noise with the help of the tissue that is disposable and avoiding contact with the year mouth and nose can help in the prevention. The world has shock due to COVID-19 because of its non-availability of drugs or vaccines. Different researchers are working on the deadly virus to make do vaccines for it. The clinical report has been named COVID, ARDS. The advancement of technology generally has a rapid effect on every field of life, whether medical or other—tools such as applications and measurement instruments, generating a huge amount of data in the digital format. Data are available immediately for mentioning through the online platforms, which will be available or accessible to anybody with a broadband connection in a matter of seconds. It has been determined that artificial intelligence is considered a promising result in the Healthcare sector as it helps analyze decision-making and data. It has been determined that COVID-19 has generally affected more than a hundred countries. Moreover, it has been determined that people worldwide are vulnerable to different consequences in the future. It is quite important in developing the control system that will result in the detection of the coronavirus. One of the main solutions and controlling the current problem is diagnosing the disease that can be easily done through artificial intelligence tools. The occurrence of the COVID-19 has generally shocked the world as there is no availability of drugs or vaccines. Moreover, it has been determined that various researchers are mainly working to win over COVID-19 viruses. Various features like IDF and TF are extracted from the clinical reports. The machine learning algorithm is generally used in the classification of clinical reports in two different classes.

Acknowledgment

This work is partly supported by VC Research (VCR 0000155) for Prof Chang. The second co-author has obtained permissions for use of images before the completion of this research on the behalf of the team.

References

Agbehadji, I. E., Awuzie, B. O., Ngowi, A. B., & Millham, R. C. (2020). Review of big data analytics, artificial intelligence and nature-inspired computing models towards accurate detection of COVID-19 pandemic cases and contact tracing. *International Journal of Environmental Research and Public Health*, *17*(15), 5330.

Balachandran, B. M., & Prasad, S. (2017). Challenges and benefits of deploying big data analytics in the cloud for business intelligence. *Procedia Computer Science*, *112*, 1112–1122.

Chu, D. K., Akl, E. A., Duda, S., Solo, K., Yaacoub, S., Schünemann, H. J., et al. (2020). Physical distancing, face masks, and eye protection to prevent person-to-person transmission of SARS-CoV-2 and COVID-19: A systematic review and meta-analysis. *The Lancet*, *395*(10242), 1973–1987.

Corsi, A., de Souza, F. F., Pagani, R. N., & Kovaleski, J. L. (2020). Big data analytics as a tool for fighting pandemics: A systematic review of the literature. *Journal of Ambient Intelligence and Humanized Computing*, 1–18. https://doi.org/10.1007/s12652-020-02617-4.

Elgendy, N., & Elragal, A. (2014, July). Big data analytics: A literature review paper. In *Industrial conference on data mining* (pp. 214–227). Cham: Springer.

Haleem, A., Javaid, M., Khan, I. H., & Vaishya, R. (2020). Significant applications of big data in COVID-19 pandemic. *Indian Journal of Orthopaedics*, *54*, 526–528.

Jia, Q., Guo, Y., Wang, G., & Barnes, S. J. (2020). Big data analytics in the fight against major public health incidents (including COVID-19): A conceptual framework. *International Journal of Environmental Research and Public Health*, *17*(17), 6161.

Lalmuanawma, S., Hussain, J., & Chhakchhuak, L. (2020). Applications of machine learning and artificial intelligence for Covid-19 (SARS-CoV-2) pandemic: A review. *Chaos, Solitons & Fractals*, *139*, 110059.

Ram, J., Zhang, C., & Koronios, A. (2016). The implications of big data analytics on business intelligence: A qualitative study in China. *Procedia Computer Science*, *87*, 221–226.

Sheng, J., Amankwah-Amoah, J., Khan, Z., & Wang, X. (2020). COVID-19 pandemic in the new era of big data analytics: Methodological innovations and future research directions. *British Journal of Management*. https://doi.org/10.1111/1467-8551.12441.
Shorten, C., Khoshgoftaar, T. M., & Furht, B. (2021). Deep learning applications for COVID-19. *Journal of Big Data*, *8*(1), 1–54.
Siam, M., Banna, H., Nishat, N. H., Ahmed, A., & Hossain, M. S. (2020). Stopping the COVID-19 pandemic: A review on the advances of diagnosis, treatment, and control measures. *Journal of Pathogens*, *2020*, 1–12.
Syeda, H. B., Syed, M., Sexton, K. W., Syed, S., Begum, S., Syed, F., et al. (2021). Role of machine learning techniques to tackle the COVID-19 crisis: Systematic review. *JMIR Medical Informatics*, *9*(1), e23811.
Tang, Y. W., Schmitz, J. E., Persing, D. H., & Stratton, C. W. (2020). Laboratory diagnosis of COVID-19: Current issues and challenges. *Journal of Clinical Microbiology*, *58*(6), e00512–20.
Tsai, C. W., Lai, C. F., Chao, H. C., & Vasilakos, A. V. (2015). Big data analytics: A survey. *Journal of Big Data*, *2*(1), 1–32.
Verma, S., & Gustafsson, A. (2020). Investigating the emerging COVID-19 research trends in the field of business and management: A bibliometric analysis approach. *Journal of Business Research*, *118*, 253–261.
Wang, L., & Alexander, C. A. (2016). Machine learning in big data. *International Journal of Mathematical, Engineering and Management Sciences*, *1*(2), 52–61.

CHAPTER 3

Designing a conceptual model in the artificial intelligence environment for the health care sector

N. Raghavendra Rao
FINAIT Consultancy Services, Chennai, India

1 Introduction

Hospitals across the globe have a reasonably good information system in managing the internal administrative and medical procedural process for their patients. While providing health care services to their patients, a large amount of data is generated. This data will be mainly about the diagnosis and treatment suggested with the prescriptions of the drugs by the doctors to their patients. Most of the hospitals do not make use of the above data for medical research purposes. Over some time, their computer systems will become slow. It is due to the huge data stored in their systems. Generally, hospitals transfer the data from their systems either to external devices or remove the data from their existing systems. Data stored in the external devices are not generally referred to for any purpose by the hospitals (McConnell, 2020).

2 Background

For years research in medicine was related closely to natural sciences. This research was mainly related to developing new drugs. Rapid advancements in telecommunication technology enabled the healthcare sector to develop Telemedicine for the benefit of patients across the globe.

It will be interesting to recall the approach followed by Middlesex University in the UK when telecommunication was becoming popular. Middle Sex University has created a central disease management system. Asthma patients and those with chronic obstructive pulmonary disease would use a portable monitoring device to record their breathing patterns up to four

Novel AI and Data Science Advancements for Sustainability in the Era of COVID-19
https://doi.org/10.1016/B978-0-323-90054-6.00004-0

times a day in the comfort of their homes. The data was sent via modem and telephone lines to the central disease management system. It was processed and results were sent to the patient's doctor using the cable and wireless secure internet way. This system would record the date and time, temperature, and humidity measures critical for analyzing the health of asthma patients surrounding such as air pollution and quality of the air. This information would assist in providing the correct treatment and diagnosis for individual patients. The patients' symptoms and use of medication, as well as their lung function data, were recorded in the central disease management. This disease management system would contain two parts of data in respect of asthma patients. One part of data would be in respect of patient's data about date and time, temperature and humidity measurements. The second part of the data would relate to environmental conditions, air pollution, and quality. This approach facilitated the doctors to treat their patients.

One more interesting aspect is that signal processing has helped electronic engineers develop electroencephalogram (EEG) and Electrocardiogram (ECG). These devices are meant to view urgent medical reports. Magnetic Resonance Imaging (MRI) scans and Positron Emission Tomography (MRI) scans are the tools made use by the medical fraternity for diagnosis.

In the present scenario, the concept of artificial intelligence is gaining importance in many disciplines. Medical science is no exception to make use of this concept for healthcare issues. This concept provides a lot of scope in analyzing the huge volume of data in the health care sector for medical research.

This analysis provides an insight into the procedures followed by the doctors in different hospitals.

Further, this concept and the collaborative concepts in information and communication technology also facilitate the design of a new drug or enhance the features in the existing drug.

3 Literature review

The authors (Abdel-Basset, Chang, & Nabeeh, 2021) explain in their paper the importance of Disruptive technologies for covid-19 analysis. Disruptive technologies considered for covid-19 analysis are AI, Industry 4.0, IoT, Big Data, Virtual Technology, IoMT, 5G, and Autonomous Robot, and Clock Chain. The proposed framework suggests restricting covid-19 breakouts.

Further, it ensures the safety of the health workers and maintains patient's physical and psychological health care.

The authors (Abdel-Basset, Chang, & Mohamed, 2020) in their paper explain the significance of X-ray images to identify the infection with covid-19. In this context, they have explained and compared five algorithms. They are (1) Whale Optimization Algorithm (WOA), (2) Fire Fly Algorithm (FFA), (3) Harris-hawks Algorithm (HHA), (4) Salp Swarm Algorithm (SSA), and (5) Standard SMA. The performance of these algorithms is clearly explained.

The authors (Abdel-Basset, Chang, Hawash, Chakraborty, & Ryan, 2021) stress the importance of deep learning for semisupervised few-shot segmentation covid-19 infection. The key challenge of this study is to provide accurate covid-19 infection from a limited number of annotated instances.

4 Approach suggested for designing a conceptual model

A huge volume of data related to healthcare issues is available in the healthcare sector. This data can be made use for analysis by the medical fraternity. The medical fraternity who have research bent of mind can make use of the data for medical research. This chapter suggests a process in creating a conceptual model for analysis and developing a drug for the benefit of patients. This chapter is divided into 10 sections. They explain the process of creating a conceptual model. The 10 sections are (1) Selection of concepts in information and communication technology, (2) Databases related to diseases, digital image and virus taxonomy, (3) Role of a core team, (4) Overview of Viruses, (5) Case illustration based on the healthcare sector in India, (6) Machine learning approach for analyzing medical data, (7) Academic research approach, (8) Developing a drug, (9) Discussion, and (10) Conclusion.

5 Selection of concepts in information and communication technology

There are advanced features in the many concepts in information and communication technology. The selection of the concepts depends on the requirements of a project. The concepts such as artificial intelligence, Machine learning, data warehouse, virtual reality, and cloud computing are appropriate for creating a conceptual model for the use of medical data

(Christensen, Grossman, & Hwang, 2008). Besides international classification of diseases is also explained. The scenario of the healthcare sector in India is explained for case illustration purposes.

The concept of machine learning is applied for analyzing medical data. The concept of virtual reality is used for discovering a drug based on the analyzed medical data from the machine learning approach. Salient features of the concepts are briefly explained. Role of the core team is highlighted.

International Classification of Diseases is useful for labeling medical data in the database is also explained. The scenario of the healthcare sector in India is explained for case illustration purposes.

5.1 Artificial intelligence

Artificial intelligence is a subfield of computer science aimed at the development of computer systems capable of doing things that are normally done by people, in particular things associated with people acting intelligently (Janikiram, Sarukesi, & Gopalakrishna, 2017).

5.2 Role of artificial intelligence

Artificial intelligence is a collection of concepts and ideas related to the development of an intelligent system. These concepts and ideas may be developed with different ideas and be applied to different domains (Patterson, 2017).

5.3 Machine learning

Artificial intelligence is the overall category that includes machine learning and natural language processing. Machine learning enables a computer system to learn from data rather than through explicit programming. There are four main subsets of artificial intelligence. They are (1) supervised learning, (2) unsupervised learning, (3) reinforcement learning, and (4) deep learning in neural networks.

5.4 Algorithms

Algorithms are a set of instructions for a computer system to interact, manipulate, and transform the data. As the machine learning algorithm is exposed to more data, it can create an increasingly accurate algorithm.

5.5 Data warehouse

A data warehouse is the central storage of data that is extracted either from a historical database or from the operational database. The data in a data warehouse are subject-oriented, nonvolatile, and historical. So data warehouse tends to contain extremely large data sets (Ponniah, 2012). It may be inferred that the purpose of the data warehouse is (1) to slice and dice through data, (2) to ensure the past data is stored accurately, (3) to provide one version of data, (4) to operate for an analytical process, and (5) to support the decision process.

5.6 Virtual reality

Virtual reality is a way of creating a three-dimensional image of an object or scene (Linowes, 2015). It facilitates a user to move through or around the image. Virtual reality imitates the way the real object or scene looks and changes. Information system helps to use information in databases to stimulate. The line dividing stimulated tasks and their real-world counterparts are very thin. The computer system uses sensory inputs such as vision, hearing, feeling, and sensation of motion.

5.7 Cloud computing

Cloud computing is a concept generally defined as a group of scalable and virtualized resources that make use of the internet to provide on-demand services to end-users (Thomas, 2014). Cloud computing provides four types of environments to end-users. They are (1) Private Cloud, (2) Community Cloud, (3) Public Cloud, and (4) Hybrid Cloud. The main use of cloud computing is for the storage of data. Data is stored in multiple servers rather than in the dedicated servers used in the traditional network data storage system. An end-user sees it as a virtual space carved out of the cloud.

6 Databases related to classification of diseases, digital image code, and viruses taxonomy

The following paragraphs provide an overview of the standards already established by different international agencies.

6.1 International Classification of Diseases (ICD)

ICD is a specific standard for medical terminologies. The ICD has become the international standard for diagnostic classification for all general epidemiological and many health management purposes. It may be noted that epidemiology is a branch of medical science that deals with the incidence, distribution, and control of disease in the population.

It is also about the sum of the factors controlling the presence or absence of a disease or pathogen. ICD-10 uses alphanumeric codes with alphabetic codes with an alphabetic character in the position. ICD-10 contains the classification of diseases. Similarly, medical images also have codes. These codes are classified in DICOM (Digital Imaging and Communications in Medicine).

6.2 Digital Imaging and Communications in Medicine (DICOM)

DICOM is a standard for exchanging medical images. Medical images comply with the DICOM format.

6.3 International Committee on the Taxonomy of Viruses (ICTV)

ICTV is given the responsibility of taking charge of the task of developing, refining, and maintaining a universal virus taxonomy.

7 Role of core team

The term cognitive limit indicates that an individual's problem-solving capability limits when a wide range of diverse information and knowledge required. Pooling several individuals may help in solving a problem. There is every likelihood hood of the problem of coordination and communication may arise in working groups. In computerized systems, it will be very easy for people to have access to a vast amount of stored information for analysis in suggesting a solution to a problem.

It will be better to form a core team for creating a conceptual model in the healthcare sector. The core team should consist of an experienced medical fraternity and software professionals. Software professionals in the core team need to make conscious efforts to understand the medical terms and medical procedures followed by the hospitals. Similarly, the medical fraternity should take an interest in understanding the features of information and communication technology suggested by the software professionals. The interdisciplinary approach adopted by the team members will be useful in designing a conceptual model for analyzing the medical data and finding an effective solution for the healthcare issues (Galea, 2019).

7.1 Medical research activities

Any research needs a large volume of data for research. There is no dearth of medical data in India (Vijayrani, 2013). It is general practice in many hospitals in India to optimize the use of a database that contains the current year's data only. Earlier year's data are either transferred to the external device as a backup, or some hospitals delete the earlier year's data from their systems.

In the interest of humanity data from the various healthcare providers are to be transferred to a centralized database (Lals, 2018). The concept of a data warehouse can be used for centralized database purposes (Kraynak, 2017). Moreover, every hospital cannot afford to have its research department for the analysis of medical data.

7.2 Virtual medical research center

A virtual research center can be created by a corporate hospital or a group of corporate hospitals. The medical fraternity and the professionals in other health service providers who have a research bent of mind can involve themselves in research activity. The advantage is that they need not move out from their workplace and at the same time, their professional work is also not affected. Another advantage is there will be a good opportunity for everyone who lives and work across India. It is needless to say that a cloud computing environment will enable the virtual medical research center to work effectively (Wulfovich, 2019). Medical research scholars can suggest the composition of a drug for a new drug or the new composition of a drug for an existing drug. The process of developing a drug is given in the following paragraphs.

8 Overview of viruses

A virus consists of two or three parts. Genes are made from either DNA OR RNA long molecules that carry genetic information. Protein coat that protects the genes in some viruses and envelope of fat. Viruses vary in shape from simple helical and icosahedral to more complex structures.

8.1 Viruses

It is said that over 380 trillion viruses are inhabiting the human body. It is collectively known as the human virus. But these viruses are not the dangerous ones commonly heard about, like those that cause the flu or the common cold. The most sinister infections are Ebola or dengue.

8.2 Spreading vectors

Viruses spread in many ways. Many viruses are very specific as to which host species or tissue they attack. Plant viruses often spread from plant to plant by insects and other organisms known as vectors. Some viruses of animals, including humans, are spread by exposure to infected bodily fluids.

Viruses such as influenza are spread through the air by droplets of moisture when people cough or sneeze. Viruses such as nor viruses are transmitted by the facial-oral route, including the contamination of hands, flood, or water.

8.3 Human immunodeficiency viruses

Dengue viruses are spread by blood-sucking insects. Rotavirus is often spread by direct contact with infected children. Antibiotics do not affect viruses. But antiviral drugs have been developed to treat life-threatening infections, Vaccines that produce lifelong immunity can prevent some viral infections.

8.4 Role of immune system

The immune system protects the human body against disease or other potentially damaging foreign bodies. When the immune system functions properly, it will identify a variety of threats such as viruses, bacteria, and parasites and attacks them. It will also distinguish them from the human body's healthy tissue.

8.5 Parts of immune system

Bacterial and viral infections can cause symptoms such as coughing, sneezing, fever: inflammation, vomiting, diarrhea, fatigue, and cramping. The immune system in the human body tries to get rid of the above infections (Boslaugh, 2013).

The main parts of the immune system are (1) White Blood Cells, (2) Antibodies, (3) Complement system, (4) Lymphatic, (5) Spleen, (6) Thymus, and (7) Bone Marrow. These are the parts of the immune system in the human body that actively fight infection (Fletcher, 2020).

8.6 Characteristics of immune system

The immune system in the human body is made up of special organs, cells, and chemicals that fight infection (microbes). The main parts of the immune system are white blood cells, the lymphatic system, the spleen, the thymus, and the bone marrow. These are the parts in the human body's immune system that actively fight infection. The immune system keeps a record of every microbe it has ever defeated in the types of white blood cells. These are otherwise known as memory cells. This means that it can recognize and destroy the microbe quickly. Suppose it enters the human body again before it can multiply and makes a person feel sick. Some infections like the flu and the common cold have to be fought many times. It is because so many different viruses or strains of the same type of viruses can cause these illnesses from other viruses. It means that it does not give the human body immunity against others.

8.6.1 *White blood cells*

White blood cells are the key players in the human body's immune system. They are made in the bone marrow and part of the lymphatic system. White blood cells move through blood and tissues throughout the human body, looking for microbes such as bacteria, viruses, parasites, and fungi. Once they find them, they launch an immune attack.

8.6.2 *Antibodies*

Antibodies help the human body to fight microbes or toxins (poisons), they produce. They do this by recognizing substances called antigens on the surface of the microbe. Or in the chemicals they produce. Then they mark the microbe or toxin as being foreign. The antibodies mark these antigens for destruction. There are many cells, proteins, and chemicals involved in this attack.

8.6.3 *Complement system*

The complement system is made up of proteins whose actions complement the work done by antibodies.

8.6.4 *Lymphatic system*

The lymphatic system is a network of tubes through the human body. The main role of the lymphatic system is to manage the fluid levels in the human body. Another important function that reacts to bacteria and deals with cancer cells. Otherwise, it will result in diseases or disorders. It also absorbs some of the fats in one's diet from the intestine.

8.6.5 Spleen

The spleen is a block of filtering organs that remove microbes and destroys old or damaged red blood cells. It also makes disease-fighting components in the immune system.

8.6.6 Bone marrow

Bone marrow is the spongy tissue found inside the bones of the human body. It produces the red blood cells in the human body to carry oxygen.

8.6.7 Thymus

The thymus filters and monitors one's blood content. It produces white blood cells.

8.7 Common disorders of the immune system

It is common for people to have an over or underactive immune system. Over activity of the human system can take many forms, including allergenic diseases where the immune system makes a strong response to allergies. They include allergy to food, medicines, sinus, and asthma. Autoimmune diseases range from common to rare.

Underactivity of the immune system is also called immunodeficiency. Examples of these conditions include primary immunodeficiency diseases such as common variable immunodeficiency (CVID). In some cases, it is linked to severe combined immunodeficiency (SCID) and complement deficiencies arise as a medical treatment.

An underactive immune system does not function correctly and makes people vulnerable to infections. It can be life-threatening in some cases.

8.8 Types of immunity

Innate immunity: People are all born with some level of immunity to resist viruses from entering the human body.

Adaptive immunity: It is also known as adaptive immunity. This protects from pathogens developing as people go through life.

Passive immunity: This type of immunity is borrowed from another source. Generally, it does not last indefinitely.

8.9 Signs of the weakened immune system

The following are considered signs of a weakened immune system (22).

1. One's stress level is high
2. One has always cold
3. One has lots of tummy troubles
4. Wounds are slow to heal
5. One has frequent infections
6. One feels tired all the time

9 Covid-19

Coronavirus is a large family of virus which may cause illness in animals or humans (Hogg, 2005). In humans, several coronaviruses are known to cause respiratory infections ranging from the common cold to more severe diseases such as middle ease respiratory syndrome (MERS) and severe acute respiratory syndrome (SARS). The most recently discovered coronavirus causes coronavirus disease (COVID 19).

The most common symptoms of COVID-19 are fever, dry cough, and tiredness. Some patients may have aches and pains, nasal congestion, sore throat, or diarrhea, and those with underlying medical problems like high blood pressure, heart and lung problems, diabetes, or cancer, are at higher risk of developing a serious illness. People can catch COVID-19 from others who have the virus. The disease spreads primarily from person to person through small droplets/respiratory droplets from the nose or mouth, which are expelled when a person with COVID-19 coughs, sneezes, or speaks. These droplets are relatively heavy, do not travel far, and quickly sink to the ground. People can catch COVID-19 if they breathe in these droplets from a person infected with the virus. These droplets can land on objects and surfaces around the person, such as tables, doorknobs, and handrails. People can become infected by touching these objects or surfaces, then touching their eyes, nose, or mouth (Kudakasseril, 2020).

9.1 Classification of Covid-19 symptoms

Covid-19 affects different people in different ways. Most infected people will develop mild to moderate illness and recover without hospitalization.

A. *Most common symptoms*:

Fever, dry cough, and tiredness.

B. *Less common symptoms*:

Aches and pains, sore throat, diarrhea, conjunctivitis, headache, loss of taste or smell, a rash on the skin, or discoloration of fingers or toes.

C. *Serious symptoms*:

Difficulty in breathing or shortness of breath, chest pain or pressure loss speech or movements.

D. *Asymptomatic covid-19*:

An asymptomatic laboratory-confirmed case is a person infected with covid-19 who does not develop symptoms.

E. *Herd immunity*:

When most of a population is immune to an infectious disease, this provides indirect protection or herd protection. This is also called herd immunity.

F. *Immune system*:

When the cells of the immune system in a human body work smoothly, a human being does not feel ill. Once it stops working smoothly or properly because it is weak or cannot fight particularly aggressive germs, one falls ill.

G. *Epidemiology*:

Epidemiology is studying and analyzing the distribution patterns and determents of health and disease conditions in a defined population. It is considered as the cornerstone of public health. It also provides evidence-based practice in identifying risk factors for disease and targets for preventive health care.

10 Case illustration based on the healthcare sector in India

There are four types of hospitals that provide healthcare services to various types of patients in India (Govil & Purohit, 2011). They are:

(1) Hospitals managed by the corporate sector
(2) Hospitals managed by the Government at the state and central government levels
(3) Hospitals managed by a group of Trustee, and
(4) Hospitals managed by an Individual Doctor or a Group of Doctors.

In Addition to the above, some specific specialists have their consulting centers for providing services in the area of their specialization. There are also many clinical testing laboratories and diagnostic centers.

The above service providers for health care issues are having information systems for their professional work. These information systems are designed and developed by software professionals based on the requirements provided by them. The healthcare service provider's information systems contain the diagnosis and the treatment suggested to their patients. Their systems also contain the drugs prescribed and reactions of the drugs on their patients and alternative drugs prescribed to the patients who have an allergy to a particular composition in certain drugs.

Medical procedures followed at the respective hospitals are also stored in their system. Information about the clinical test results of each patient is available in their systems. Reports about x-ray and scanning of a particular part of the human body with the images are available at the respective diagnostic centers.

10.1 Data in the centralized database

The centralized data warehouse should contain the following data from the various sources of the healthcare sector in India. This data warehouse can be stored in a private clouding environment.

(1) Details of data such as patients' profession and living in which part of India.
(2) Habits and all health issues related to the patients.
(3) Procedures followed tests conducted and results of the tests.
(4) Treatment, suggested medicines prescribed and the reaction of the medicines experienced by the patients and.
(5) The data relating to the patients of similar complaints from the various hospitals.

11 Machine learning approach for analyzing the medical data

In the present covid19 scenario, software professionals who are developing machine learning algorithm, particularly for the covid19 they need to understand the following areas such as epidemiology and coronavirus.

The centralized data warehouse will contain the data of other diseases from the various sources of the health care service providers. The machine learning methods such as supervised learning and unsupervised learning convert the data into a structured way and provide insight into the analysis of health care issues.

Supervised learning and unsupervised learning approaches (Hurwitz & Kirsch, 2018) can be applied to the centralized data warehouse mentioned in the case illustration of this chapter.

11.1 Supervised learning

Supervised learning typically begins with an established set of data and a certain understanding of what is classified. Before applying this technique, the medical data available in the centralized database needs to be classified with ICD codes. Supervised learning is applied to find the patterns in data. The patterns facilitate the analytical process. The ICD codes have the characteristics that define the data. This helps to understand the correlation between the side effects experienced by the different patients for the same com of the various viruses.

11.2 Unsupervised learning

Unsupervised learning is best suited when the health issue requires a massive amount of data that is unlabelled. This approach is more useful for the centralized data warehouse mentioned in the case illustration of this chapter. The centralized data warehouse base contains the data from various sources in the healthcare sector in India. Service providers do not follow the International Code of diseases. Understanding the meaning behind this data requires algorithms that can begin to understand the meaning based on the patterns or clusters it finds. The unsupervised process of analyzing data without human intervention.

Unsupervised learning algorithms segment data into groups of clusters or groups of features. The unlabelled data creates the parameter values and classification of the data. In essence, this process adds labels to the data. Besides, ICD codes can be made use of labeling so that it becomes supervised learning. Unsupervised learning can determine the outcome when there is a massive amount of data.

In healthcare collecting, huge amounts of data about a specific disease can help medical practitioners to gain insights into the patterns of symptoms and relate those to outcomes from patients. It will take too much time to label all the data sources associated with a disease, such data sets. Therefore an unsupervised learning approach can help to determine outcomes more quickly than a supervised learning approach.

12 Indian environment

Generally, a person has a fever, that person takes aspirin or a similar tablet.

If there is a dry cough, cough syrup is taken. When a person feels tiredness, that person takes off from the regular activity and relaxes. If these symptoms persist, then consult a doctor. On the advice of the doctor, medicine is prescribed by the doctor the patient takes. In the present scenario, this is considered a mild symptom in the covid-19 virus.

The data for the mild symptoms and treatment prescribed are available in the centralized data warehouse in the system. This data can be made use of for analysis. This analysis will provide for the classification of the number of patients who had the above symptoms, their place of living, and other relevant information.

This classified information will be useful for comparison and analysis before the pandemic and during the pandemic. This will facilitate applying the concept of epidemiology for adopting a workable approach for managing the pandemic scenario.

12.1 Less common symptoms

These symptoms depend upon one's lifestyle of the patients. Each symptom and the combination of symptoms under this category can be identified in the centralized data warehouse for analysis and comparison before and present the covid-19 scenario. This approach will provide to conduct a serious study of these symptoms. The result of this study will provide scope for adopting a workable solution to manage these symptoms.

12.2 Serious symptoms

These symptoms depend upon one's health condition of a patient. The patients who have had these symptoms in the present pandemic need to be studied and analyzed for developing a drug.

12.3 Rule-based machine learning

Rule-based machine learning algorithms use relational rules to describe data. A rule-based system can be contrasted from machine learning systems that create a model that can be generally applied to incoming data. In the abstract rule-based systems are very easy to understand. As the system encounters more data and is trained, hundreds of exemptions to the rules might likely emerge. It is important to be careful when creating a rule-based approach that it does not become so complicated that it loses its transparency.

Artificial Immune Systems are a class of computationally intelligent, rule-based machine learning systems inspired by the principles and processes of the vertebrate immune system (Zhan & Hou, 2014). The algorithms are typically modeled after the characteristics of the immune system of learning and memory for use in solving health issues. The common techniques are adopted by specific immunological theories that explain the function and behavior of the mammalian adaptive immune system. Originally the artificial immune systems were set to find efficient abstractions of processes found in the immune system. It is now being applied in modeling the biological process and applying immune algorithms to bioinformatics problems (Read, Timmis, & Andrews, 2012). Characteristics of the immune system can be made use of for designing an artificial immune system to understand the working of the immune system in a human body.

12.4 Algorithms

A rule-based approach under artificial intelligence algorithms can be developed for identifying the symptoms related to less common and severe from data available in the centralized data warehouse. Three types of analysis can be carried out. They are (1) before the covid-19 virus started spreading, (2) during the pandemic, and (3) new types of health issues arose during the lockdown period.

New health issues such as changes in sugar levels and depression were recorded in the centralized data warehouse. An in-depth analysis of the new health issues will facilitate a workable approach to be developed in managing these issues and the probable health issues that are likely to arise. The concept of epidemiology can be used to frame a procedure to be followed in handling these issues.

13 Academic research approach

The centralized data warehouse mentioned in the case illustration referred to in this chapter will contain the data about the patients who have suffered the various viruses given in the overview of the viruses and diseases. It will contain the data related to the drugs prescribed to the patients. It will also have data containing the information related to the number of patients cured by the drugs prescribed to them. It will have information about the patients.

Suffered side effects due to the drugs did agree with them.

If one closely observes, some symptoms mentioned under covid19 are prevalent in other types of viruses mentioned in the paragraph of the overview of viruses. In those cases, the drugs prescribed helped the patients to get cured.

The following aspects can be analyzed from the results obtained by adopting the supervised and unsupervised approach of a machine learning concept.

1. Number of patients suffered the various types of viruses and diseases in the different parts of India.
2. Level of immunity among the patients who have been treated by the various hospitals in India.
3. Comparision of the common symptoms of the various viruses for academic analysis.
4. Data related to covid19 need to be analyzed for developing a drug for this virus.

The research team members of the virtual research center can suggest the composition of a drug based on their analysis and experience for the covid19 virus. They can also suggest a new composition of an existing drug that did not agree with the patients for the virus other than covid19. The process of developing a drug is explained below.

14 Developing a drug

A drug is a molecule that interacts with the target biological molecule in the human body. It is through such interaction that it triggers a physiological effect. The target molecules are usually proteins. Drugs can be beneficial or harmful depending on their effect (Jain Vivek, 2020). The aim of discovering a drug is for specific beneficial effects to treat diseases in human beings.

Discovering a drug can be arrived at by two methods. The methods are empirical and rational. The empirical method is a blind or loose method. Thousands of chemical compounds are tested on the disease without even knowing the target on which the drug acts and the mechanism of action. The rational method starts from the clear knowledge of the target as well as the mechanism by which it is to be attacked. Drugs act either to stimulate or block the activity of the target protein.

14.1 Role of bioinformatics in drug development

Bioinformatics is the storage, manipulation, and analysis of biological information by making use of information technology. Bioinformatics is an essential infrastructure underpinning biological research (Bergeron, 2003). The adoption of the bio informatics-based approach to drug discovery provides an important advantage in the rational approach.

14.2 Biological data

The properties that characterize a living organism (SPECIES) are based on its fundamental set of genetic information. It is important to understand the fundamental aspects of terms such as DNA, RNA, Protein and their information concerning the genome.

Different sequences of bases in DNA specify different sequences of bases in RNA. The sequence of bases in RNA specifies the sequences of amino acids in proteins. The central dogma states that DNA is transcribed into RNA, which is then translated later into protein (Ignacimuthu, 2005).

The main advantage in the bio-informatics discipline is biological information available on various websites. The databases in these websites can be classified into two types such as generalized and specialized databases. The generalized databases contain information related to DNA, Protein, or similar types. The generalized databases can again be further split into sequence databases and structured databases. Sequence databases hold the individual sequence records of either nucleotide or amino acids, or proteins. Structured databases contain the individual sequence records of biochemically solved structures of Macro-molecules (Claverie & Notredame, 2011).

Specialized databases are

(1) EST (Expressed Sequence Tags).
(2) GSS (Genome Survey Sequences).
(3) SNP (Single Nucleotide polymorphism.
(4) STS (Sequence Tag Sites).
(5) KABAT for immunology proteins and ligand for enzyme reaction legends.

These databases can be further split into three types based on the complexity of the data stored.

(1) **Primary databases:** These databases contain data in its original form from the sequences.
(2) **Secondary database:** These databases have value-added data and derived information from the primary databases.
(3) **Composite databases:** Composite databases amalgamate a variety of different primary databases structured into one. There are various software tools available to facilitate search in the above databases.

Developing or designing a drug is possible by using the information in the diverse chemical libraries and the information about biological functions stores in the above databases before starting a laboratory-based experiment. It is always possible to generate as much information as possible about potential drug and target interaction from the above databases and chemical libraries.

14.3 Role of virtual reality in drug design

The software professionals who are part of a drug discovery research project can create a simulated version of a human body with all the internal parts in the human body visible under a virtual reality environment (Chorafas Dimitris & Steinmann, 1995). They can take the data from the health care sector in respect of the reactions to the drugs for highlighting them in the simulated human body system. Various reactions such as feeling restless, screaming with pain, drowsy, not getting a sound sleep and digestive problems can be viewed in the simulated version of the human body. This will facilitate the research team members to visualize the reactions of a drug in the simulated version of a human body. Each member of the research team can analyze the visual reactions in the simulated human body system. Each research team member can suggest the various components to be used in designing a new drug. The entire research team can analyze and discuss among themselves each member's suggestion of composition for a drug. After the discussions among themselves, they can mutually ages to a particular composition for a new drug. The advantage of creating a simulated human body system under the virtual reality environment can facilitate the research team members to visualize the pain and suffering due to the side effects of a drug. The drug designed under this environment can be given to a laboratory at a pharmaceutical company to produce a drug and test it before the actual manufacturing of a drug takes place (Flynn, 2019).

15 Discussion

The concept of artificial intelligence is providing a good scope for radical changes in designing and developing conceptual models in the health care sector. Data in the health care sector in India are available as disperse elements. The concept of artificial intelligence will facilitate the compilation of these elements into meaningful patterns. These patterns will be useful for research activity among the medical fraternity. Integration of knowledge and experience of medical professionals with the active support of software professionals can design and develop effective health care models for the benefit of patients.

16 Conclusion

Many technologies tend to be applicable only in a specific sector or defined areas of operations in a particular discipline. In the case of information and communication, technology is different. The potential of information and communication technology in any sphere of activity in any discipline is more useful. The concepts in information and communication technology are the driving force in providing a good scope for an open, innovative approach for any activity under the sun. The concept of artificial intelligence, along with the concepts in collaborative technology, can facilitate the medical professionals in the health care sector to design and models for their professional work (Walshe, 2014).

References

Abdel-Basset, M., Chang, V., Hawash, H., Chakraborty, R. K., & Ryan, M. (2021). *FSS-2019 Nov: A deep learning architecture for semi-supervised few shot segmentation of Covid-19 infection: Knowledge-based system. Vol. 212* (p. 106647). Elsevier. 5th January 2021.

Abdel-Basset, M., Chang, V., & Mohamed, R. (2020). *Hsma-Woa: A hybrid novel slime mould algorithm with whale optimization algorithm for tracking the image segmentation problem of chest X-RAY images: Applied soft computing. Vol. 95* (p. 106642). Elsevier.

Abdel-Basset, M., Chang, V., & Nabeeh, N. A. (2021). *An intelligent frame work using disruptive technologies for Covid-19 analysis: Technological forecasting & social change. Vol. 163* (p. 120431). Elsevier. February 2021.

Bergeron, B. (2003). *Bioinformatics computing*. Pearson Education.

Boslaugh, S. E. (2013). *Healthcare systems around the world a comparative guide*. Sage Publishing.

Chorafas Dimitris, N., & Steinmann. (1995). *Virtual reality practical applications in business and industry*. Prentice Hall.

Christensen, C. M., Grossman, J. H., & Hwang, J. (2008). *The innovator's prescription: A disruptive solution for healthcare*. McGraw Hill.

Claverie, J. M., & Notredame, C. (2011). *Bioinformatics for dummies*. John Wiley & Sons.

Fletcher. (2020). *Clinical epidemiology the essentials*. LWW.

Flynn, S. M. (2019). *The cure that works: How to have the world best healthcare*. Regnery Publishing (A Division of Salem Medical Group).

Galea, S. (2019). *Well what we need to talk about when we talk about health*. Oxford University Press.

Govil, D., & Purohit, N. (2011). Healthcare systems in India. In H. S. Rout (Ed.), *Healthcare systems—A global survey* New Century Publications.

Hogg, S. (2005). *Essential microbiology*. John Wiley & Sons Ltd.

Hurwitz, J., & Kirsch, D. (2018). *Machine learning for dummies*. John Wiley & Sons.

Ignacimuthu. (2005). *Bio informatics*. Narsoa Publishing House.

Janikiram, V. S., Sarukesi, K., & Gopalakrishna, P. (2017). *Foundation of artificial intelligence and expert systems*. Trinity Press.

Kraynak, J. (2017). *Cloud data warehousing dummies.* John Willey & Sons.
Kudakasseril, J. (2020). *Impact of Covid19 with special emphasis on India.* Notion Press.
Lals. (2018). *Public health management principles.* CBS Publishers & Distributors.
Linowes. (2015). *Virtual reality projects.* Pack7.
McConnell, C. R. (2020). *Hospitals and healthcare systems: What they are and how they work.* Jones & Bartlett Learning.
Patterson, D. W. (2017). *Introduction to artificial intelligence & expert systems.* Pearson India Education Services.
Ponniah, P. (2012). *Data warehousing fundamentals for IT professionals.* Wiley.
Read, M., Timmis, J., & Andrews, P. S. (2012). *An introduction to a practical immune systems* (p. 278701678). www.researchgate.net/Publications.
Thomas, E. R. L. (2014). *Cloud computing concepts, technology, and architecture.* Pearson.
Vijayrani, S. (2013 December). *Economic health records—An OverView.* CSI Communication Journal.
Vivek, J. (2020). *Review of preventive & social medicine (Including bio-statistics).* Jaypee Brothers Medical Publisher.
Walshe. (2014). *Healthcare management.* McGraw-Hill.
Wulfovich, S. (2019). *Digital health entrepreneurship.* Springer.
Zhan, J., & Hou, Z. (2014). *Application of artificial immune system in structural health monitoring.* Hindawi Publishing Corporation.

CHAPTER 4

Augmented reality, virtual reality and new age technologies demand escalates amid COVID-19

Amin Gasmi[a,b] and Rachid Benlamri[c]
[a]Société Francophone de Nutrithérapie et de Nutrigénétique Appliquée, Villeurbanne, France
[b]Interuniversity Laboratory of Motor Biology, Claude Bernard University, Villeurbanne, France
[c]Department of Software Engineering, Faculty of Engineering, Lakehead University, Thunder Bay, ON, Canada

1 Introduction

Augmented reality technology is a computed image generator that reveals visually real-world images in composite sizes and views. Such technology is used by many surgeons in AR telesurgical centers to view different internal organs, especially bone reconstruction and other internal medical complications. Medical AR has a 2-way audio stream integrated into webcams to highlight with annotations and drawing the structural component of bodies in examination (Greenfield et al., 2018). AR makes it possible for health practitioners to achieve data visualization for diagnosis and treatment, which enhances efficiency, safety, and surgical treatment costs.

Medical Scientists in 1895 carried out the first medical imaging experiment to examine human X-rays, marking the starting and ending points using a well-developed ultrasound (USG) tool. Since then, much progress has been made in medical imaging, developing various imaging modalities, such as computed tomography (CT), MRI, and 2-3D dimensional image tools that show human anatomy, its functional and possible diagnosis. Most physiological therapies make use of either MRI (magnetic resonance imaging), FMRI (functional magnetic resonance imaging), or SPECT/CT (single-photon emission computed tomography). The purpose of this advanced medical imaging technology is to capture real-time data, navigate human anatomy, display microscopes, etc. (Vávra et al., 2017).

In conjunction with modern medical imaging technology, AR/VR has been heavily used in telemedicine, allowing multidisciplinary teams to collaborate and remotely monitor patients to eliminate the shortage of physicians, especially the need to visit nursing homes and hospitals to support the

Novel AI and Data Science Advancements for Sustainability in the Era of COVID-19
https://doi.org/10.1016/B978-0-323-90054-6.00005-2

fight of COVID-19. The use of AR/VR remote procedures minimizes risks of contamination and waste of PPEs. Medically, VR offers virtual illustrations of the world, which can be overlapped with AR real-world virtualization. AR/VR helps physicians to gain insight into patients' well-being and share their discoveries with other experts using mobile devices (Ferrucci et al., 2009). Virtual and Augmented Reality works with a head-mounted device (HMD) that perceives human body systems and identifies human brains. The light from AR/VR can propel real-life appearances of the brain in two visual processing forms known as monocular and binocular processes.

The monocular process captures a single image at a time, while binoculars capture two or more images. A binocular process is publicly accepted due to its strong effects and in-depth data structure used to create stereopsis via dual-image illustration. Binocular also identified human anatomy using 3D immersive sensor tools (IWAR'99, 1999). There are different types of AR/VR projection, but perspective projection specifically displays smaller objects, as the projecting device used can be placed far away from the object. The occlusion AR/VR projection involves the display of an object in front of other blocked objects. AR/VR comes in different sizes, shades, structures and motion parallax (Abdel-Basset, Chang, & Mohamed, 2020).

Before Hollander and Carr (2020) received his first tele-sphere HMD mask patent right, Sutherland (1965) used the traditional half-developed VR to perceive real-world environments. Hence, the Morton HMD mask has a reliable 3D immersed simulator and enabled stereoscopic sound display known as Sensorama. Though, the clinical use of VR commenced in the mid-1990s when experts attempted to use it for surgical training, planning, and therapeutic rehabilitation (Tang, Owen, Biocca, & Mou, 2002). Azuma et al. (2001) describe how AR coordinates real-virtual objects by running a collaborative real-time object using cameras to track and observe the environment. AR visual display component shows a clear definition of the physical world and comes in two types: see-through optical (OST) and see-through video (VST). Most Augmented Reality VST shows virtual images in the real format, while Augmented Reality OST uses the reflection of light that passes through the human retina and lens to offer real-world visualization.

In 1993, Loomis Corporation introduced the first medical AR GPS that senses digital spatial and audio information. Fuchs et al. (1996) used a patient undergoing biopsy to demonstrate the benefit of AR. Some software developers later developed Hololens SDK and ARToolkit to advance the clinical use of AR. Google recently explored its Google Glass AR software and

Hololens to investigate AR benefits in controlling infections (Wong et al., 2020). The above-mentioned examples represent just a few use cases of AR/VR technology in medicine. In this paper will describe other applications of AR/VR to fight COVID-19 pandemic.

The reminder of this paper is organized as follows: Section 1 introduces the use of AR/VR in COVID-19 pandemic. Section 3 provides technical background of AR/VR technology and a review of the literature. Section 4 explains the demand for AR/VR technology due to COVID-19 pandemic. Finally, conclusions are drawn in Section 5.

The image below illustrates how SARS Cov-2 attacks the body from the lungs to other parts of the body (Fig. 1).

2 Updates ways to minimize infections

2.1 Vaccination for controlling the spread of Covid-19

World health organization has established cold chain vaccines equipment to achieve safe-injection of the Covid-19 vaccine without compromising drug potency. AR/VR can be used to estimate vaccine or amount of injection equipment needed to serve a specific location or targeted population.

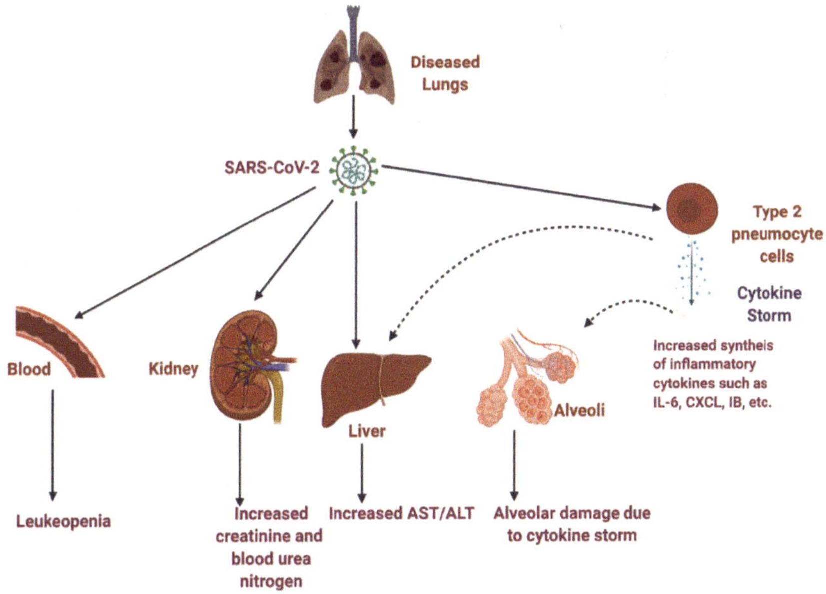

Fig. 1 SARS Cov-2 attacks the body from the lungs to other parts of the body.

AR/VR monitoring tool may eliminate waste multiplication factors (WMF) experienced during previous virus vaccinations.

XR trajectory innovative tool creates a stimulus-response for salvaging the tourist industry from being disadvantaged by the Covid-19 pandemic. Recent research proposes that the "XR trajectory" can be utilized to reshape the collapsing tourist industry (Light & Brown, 2020).

The introduction of vaccines won't stop social distancing and covid-19 monitoring which makes AR/VR usable for maintaining consistent safety measures. VR was found useful for remote medical assistance, drug discoveries and remote training and collaboration during the pandemic. Recently, updated research proves that VR helps to visualize viral proteins to discover how it works and how its inhibited enzymes can suit vaccine development. VR has found its place in Phygital therapies (treatment of psychosomatic surgery) which involves physical therapy, rehabilitation, patient distractive treatment, and correction of health disorders. AR can minimize the spread of covid-19 by offering a consistent review of the spread whereby limiting the spread to a specific location and allows communities, institutions, schools and workplaces to reopen. AR/VR boost dependent on online classes and virtual world whereby limiting human to human contact, contamination, and isolation of infected individuals.

Baidu healthcare in china launched a new app segment that works as an augmented/virtual reality asset for conducting covid-19 vaccination, online vaccine consultation, and appointment scheduling. Hong Kong called Anti-protense introduced AR/VR tool named "WIMI" with multimedia 3D modeling for real-time tracking, registration, intelligent interaction, and sensing. WIMI has computer-generative texts, images, 3D modeled music/videos, and other VR informative tools. It enables people to shop, communicate and socialize visually during the pandemic while minimizing the spread of the infection. WIMI is known as a holographic AR tool for performing visual presentation, 5G communication, Hologram ARSDK payment, etc. (Kwok & Koh, 2020).

2.2 Drugs and nutrients for controlling the spread of Covid-19

The Sars-Cov-2 has a profound protease (MPRO) enzyme which science relies upon to discover the correct anti-viral treatment. The replication of protease enzyme known as enzyme inhibitors causes the spread of the virus. Hence, researchers are focusing on identifying molecules causing the replication and stopping its mutation.

AR/VR 3D model structures are used for interactive molecular dynamic simulations to visualize molecules. IMD-VR tool enables the international drug discovery community to predict the leading cause of SARS-Cov-2 and produce relevant vaccines to combat the virus (Ferrucci et al., 2009).

Some nutritional and immune defense dietary component that boosts body system with micronutrients includes A, D, E, C, B6, and B12. It's necessary to adopt the use of Folate and selenium found in fresh meat and plant-based meals while consuming foods rich in Zinc and iron. Good dieting innate immune response controlled by T-cells which recognizes the antigens called cytotoxic. Cytotoxic T-cells are responsible for killing bodily infections and damaging viral cells while repairing T-helper's cells such as Th1 and Th2.

Food plants like onion, garlic, barberry, papaya, bitter orange, turmeric, fig, soybeans, licorice, wolfberry, mango, mulberry, black-cumin and black pepper; act as immunomodulation and anti-viral functional nutrients for subduing the spread of viruses. Some researchers claim that guava, pomegranate, and ginger can provide vitamins, fatty acids/oils, and flavonoid needed to minimize the spread of the virus in the body (Amirghofran, 2010).

3 Background and literature review

3D visualization modes elaborate the gap between memorizing, reasoning, and understanding of 3D structure represented as orbital molecules, geometry, and bio-molecular compounds. The future is yet to come, but the possibility that students will wear VR headsets, positioning their gaze while learning through image assimilated stereoscopic depth perception. Digital classrooms using VR/AR can be viable, but the incorporation of interactive learning in medicine to form both a static and spatial-dynamic virtual laboratory is yet to be effective. Meanwhile, App implemented Pixar's Universal Screen description (USDZ) as an object-generated workflow for chemistry students to organize files and implement chemical bonding. Also, Apple attached AR Quick Look app to the Safari browser to offer quick lectures by the enabled AR/VR 3D tool. This quick lecture has a less-distracting blank background known as 3D "USDZ" file that performs command-line activity using USDZ Python Script (Held & Hui, 2011). There are open-source options named Octave, Pymol, Meshlab, and Blender that handle different medic-scientific animated learning approaches.

Over the years, AR/VR has adaptive interfaces for molecular visualization and modeling in science and technology. In technology, AR/VR has specialized hardware (wearable) and software setup for research and education. AR applications can work on webpages, webcams, and other internet browsers, for examining the display of molecules, electron-microscopic maps, and orbital molecules; exhibited using HTML codes. For the molecule's mechanics to be exhibited in a real-time experimental observation, an interactive JavaScript will be integrated to provide virtual alternatives of the objects. Once the object is displayed in VR Computer Software for augmentation, the modeling kits will help to enhance images for spatial interaction, reactivity, and energetic modeling (Abriata, 2020).

In science, AR/VR has helped chemistry and structural biology disciplines by offering hybrid and computational virtual molecules of real 3D spaces of chemicals for proper calculations (Fjeld & Voegtli, 2002).

Kato and Billinghurst (1999) explained AR/VR for detecting and tracking objects using Ad Hoc hardware with sensors and cameras to track users and control the geographic spread of the virus by examining human hands for traces of dirt causing infections.

Sirakaya and Sirakaya (2018) suggested the use of telemedicine audiovisual technology to care for patients in rural areas while setting up low-resource technology to limit urban visit; while controlling the spread of the virus. Telemedicine can be viewed as a virtual reality that limits physical encounters or in-person visits to the clinic. Telemedicine can promote pre-post-operative consultation.

The American institution for Otolaryngology for neck and head surgery implored virtual reality utilization for post-operative patient checkups. This technique allows virtual visits using oncology surveillance placed on patients to monitor the recovery process. The virtual visit helps physicians to check patients' anatomy and pathology to see how fast they're recovering (Hildrew, 2020).

Medic-AR aids X-ray vision to see through real-world objects, some experts referred to the process as computer vision with pre-intra-post interventions. Medical experts explained oncology tumors, which can be envisioned with AR/VR before administering Oropharyngeal treatment and tumor removal surgery. AR enables tumor assessment using CT or MRI to determine the stage, possible diagnosis, and mapping of the treatment plan. Most doctors performing oral or Cranio-Maxillo-Facial surgery utilize anatomical AR visualization to increase the spatial outlook in a natural 3D form. This 3D reflects the oncology bone resection in an expansive virtual space (Fuchs et al., 1996).

Lu, Wang, Zhou, and Fu (2019) researched chronic Endovascular Pain in the limb of an aging adult. Other research used his research to insinuate the possibility of AR used in detecting the exact painful location. AR/VR tool coordinates with indoor positioning technologies such as radio frequency identification (RFID), Wi-Fi, visible light, Zigbee, and Bluetooth; can obtain Covid-19 records and monitor the control of COVID-19 during quarantine periods.

Ackerman (2020) devised the use of Ava robotics (IRobot) VR that can virtually show emergency cases and minimize threats of virus spread. High Altitude Platform (HAPs) promotes video connectivity to rural areas using Google Loons and Facebook Aquila to offer the possible solution. Covid-19 on the other hand, leads to an extensive use of virtual experience such as Google Hangout, Zoom and Skype as virtual labs, classrooms, and conferences.

Different researchers detailed their discoveries on the benefits of AR/VR in the educational environment, as well as the control of viruses and social distancing. Chang and O'Sullivan (2005) highlights that teaching and learning with AR increases the possibility of achieving higher learning objectives. Carlson et al. (2013) emphasize on how world education can be facilitated and transformed using VR/AR applications. López-Pérez, Pérez-López, Rodríguez-Ariza, and Argente-Linares (2013) stated that 21-Century tools may present great opportunities to achieve permanent transformation. Balsa-Barreiro, Vié, Morales, and Cebrián (2020) researched the increase of academic interest and participation if VR replaces traditional classes. Wong et al. (2020) explained the duty of spatial skill development in medical schools. AnimaRes 3D medical animation and interactive AR/VR focused tools can support the pharmaceutical discovery of COVID-19 vaccine through flow-simulation, spatial explanation, and heart rate reading (Chandola, 2020).

3.1 Interrelationship between AR and VR

Abowd and Mynatt (2000) defined AR as a technology that creates a real-based interface which projects real-world objects to be illustrated on a virtual device like a computer. AR can create mixed reality by showing virtual environments and augmented reality. With AR, the real object can be replaced with the visual object which enables the creation of virtual reality (Ahmad & Musilek, 2006).

Benford, Greenhalgh, Reynard, Brown, and Koleva (1998) attempted the classification of AR and considered VR using telepresence to mix real and virtual objects to establish a real alignment and integration with 3D real-time integration. Some AR has head-mounted displays (HMD) that can apply to human senses.

Market Survey revealed why AR technology outpaced VR, in regards to revenue, market penetration, and consumer expectation, because VR typically severs communication, gaming, training, and entertainment while AR has diverse attributes for advancing healthcare and medical training. Based on what survey predicted, AR will overtake VR in regards to sales and market position in a couple of years. However, both represent spatial computable innovations (Perkins, 2020).

The demand for AR outweighs that of VR because AR offers human sensor visibility of sight, sound, and touch which can be felt in all human sense organs. AR has haptic (touch) display attributes, smell (olfactory), and taste (gustatory) display component (Ishii & Ullmer, 1997) discussed AR Haptic audio sound system and addressed its aural display application with limited self-explanatory mono zero-dimensional phase. The aural display has other-dimensional phases that serve stereo-attributes (1-dimensional phase), headphones and loudspeakers (2-dimensional phase), advanced 3D simulator (aids diverse virtual environment Some haptic audios are sold as a consumer device, named Turtle Beach Ear force Headset (Chang & O'Sullivan, 2005).

The image below shows the interrelationship between AR and VR and it helps to track, detect and diagnose Covid-19 (Fig. 2).

3.2 Software and hardware components of AR/VR applications

As the world estimates the market growth of $125.19 billion from 2020 to 2024; NetTech AR company developed 3D photorealistic holograms called Aritize app. This app offers remote video training and live streaming. Microsoft Hololens established a Six-Degree of Freedom (DOF) augmented reality with HMD attributes. It has unique features like the field of views (FOV), a stereoscopic pixel resolution that demonstrates spatial audio-tech, Wi-Fi, and Bluetooth devices. HoloLens is an augmented glass built as an onboard computable general-purpose processor device with a holographic processing unit (HPU).

HoloLens enable the collaboration of hand, gaze, and voice inputs, allowing an optic wave lens to project images in front of the user's eyes.

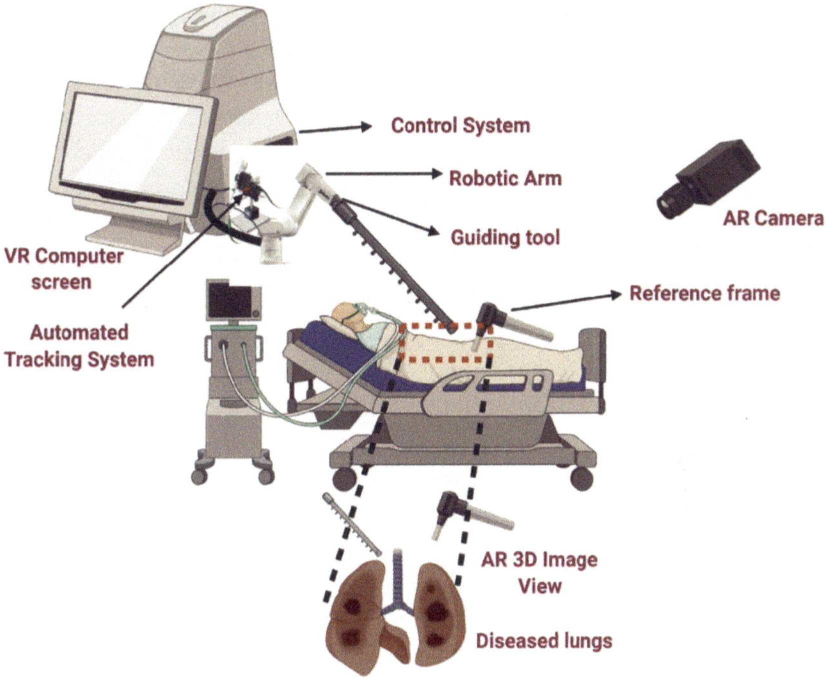

Fig. 2 AR and VR for detecting, and diagnosing Covid-19.

At present, China uses Hololens to store and track previous locations of people infected with COVID-19 by creating 3D maps using sensors to pinpoint the identified locations. It also comes with an incorporated inertial measurement unit (IMU) and sensors that reflect similar positions, rotate to other positions while allowing the head in the HMD rotation (Microsoft, 2019).

HoloLens clicker enables users to communicate with holograms by selecting, scrolling, moving and resizing applications. AR works with the Rasberry PI component necessary for connecting HoloLens to IoT using its four USB ports, quad-core ARMU8CPC, wireless LAN, Bluetooth, and Low Energy (BLE). AR integrated on IoT devices can be implemented on the Unity 3D Game Engine (Unity Technologies, 2005) to visualize real environments. To integrate AR to work with IoT for a better virtual experience, the HoloLens must work with sensors to either offer biomedical or data leveraging on IoT client-servers through messaging queuing telemetry transport (MQTT) protocols. The collaboration of AR and IoT in HoloLens is known as "Broker" which interacts with MQTT using open source MQTT python client Pahon eclipse application. Common technologies relevant for performing in-depth sensing of VR/AR include stereo vision, structured-light, and time of flight (TOF).

Stereo and structured-light interface follows triangulated models required to separate multiple capture views, using multiple sensors, light, and camera paths. The stereo vision can sometimes be passive, which helps the depth map to reproduce different pixel positions of the captured images. AR/VR structured path handles baseline image separation. The time of flight (TOF) measures the time offset and phase shift between the emitting and receptive signals. AR/VR applications normally adapt to TOF sensors through the modulated signal, measure phase shift, and depth information (Xu & Hua, 2020).

Popular depth-sensing applications used in AR/VR devices are known to evaluate depth map resolution, depth range accuracy, computation power and size, low/bright high performance, and speed (Xu & Hua, 2020). AR user interface works perfectly in unity framework, C# scripting language, and mixed reality toolkit (MRTK). The mixed reality toolkit is an open-source collection script and its components are compatible with MS HoloLens. The images below show AR/VR devices (Fig. 3).

4 Demand escalation of AR/VR due to COVID-19 pandemic

The global health resistance to viruses has by far been stretched due to the spread of COVID-19. In 2018, the global investment in AR/VR worthed \$641 million and increased to \$3.8 billion in 2020 due to COVID-19.

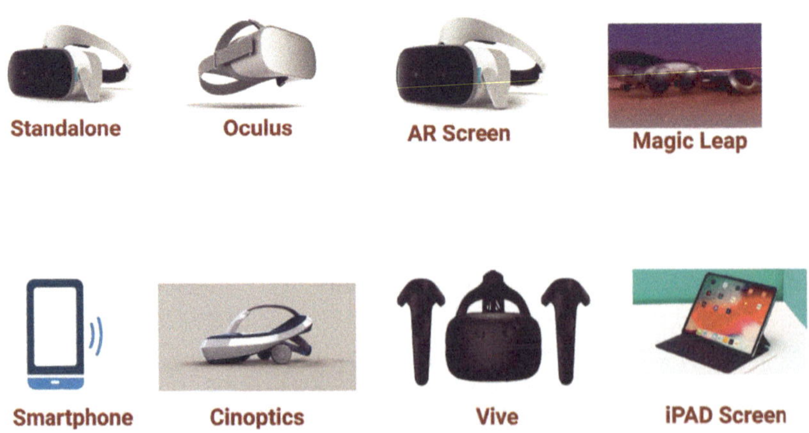

Fig. 3 AR/VR devices.

Companies manufacturing VR, AR and MR technologies in Asia confirmed a 50% increase in demand. IBM oriented their AR products to offer remote capability using an upgraded network that enables distance learning and collaboration. New emerging applications for video conferencing, webinars, remote access and emails are generally called the "Extended Reality (XR) that allows people to work from home, distance education, home fitness, private entertainment, and social distance interaction (Businesswire, 2020).

Retail stores and malls are presently using high-end AR mirrors for product visualization to prevent consumers from direct contact and virus spread. AR finds its place in cosmetics as Skywell Software used the "Cosma AR makeup app" to provide a detailed virtual outlook of products on customers. VR may influence tourism because tourists may be left with an option to explore locations with parachutes while viewing through VR applications. The real worth of the 3D depictable anatomy tool is evident in its usability to achieve multiplanar and 3D segmented image axial, computed tomography (CT), likewise magnetic-resonance image (MRI) (Gotra et al., 2017). For specialists to achieve complex Angiographic procedures, they need cone-beam rotational angiography that offers 3D segmentation of the images (Chehab, Brinjikji, Copelan, & Venkatesan, 2015).

For COVID-19, 3D and 4D illustration can be attained with an immersive holographic stereoscopic display tool; necessary for detecting virus spread throughout the body anatomy. They can also use it to study the orientation of the virus, pathology, and level of complications in detail. Held and Hui (2011) explained how trainee surgeons and medical apprentices used AR to measure surgical landscape, illustrate bodily anatomy, and understand the pathology.

On the other hand, VR applications such as a head-mounted device can provide gaze tracking, body movement observation, and heart rate examination. For surgical training, students can learn about human anatomy using VR while advancing their experience in laparoscopic and robotic surgical training using the same tool. The exposure of human anatomy during pre-surgical procedures will enhance surgical efficiency, accuracy, safety, improve and reduce operative time and minimize surgical errors.

Different VR simulators are designed to examine different types of body organs (Li & Xia, 2020) identified procedure Virtual Arthroscopy for orthopedic surgery. The insight into Arthro-VR will enable the simulation of shoulders in 3D form. VR simulators are also available for trauma vision examination, neurosurgery, and map mentor observation.

Parkhomenko Egor et al. (2018) discussed urologists' use of VR for preoperative renal anatomy examination while performing percutaneous nephrolithotomy procedure. VR has been found to alter operative approaches and reduce Fluoroscopy time, blood losses, while offering access to the hidden internal body tracts. VR is presently used by different medical scholars to perform intensive Hemodynamic observation, diseases awareness demonstration, and treatment (Papa et al., 2020).

The DEEP VR Meditative Game developed by Radboud University can serve as a bio-feed mechanism and a control system for patients with COVID-19 stress-related anxiety (Chandler, 2020). Reports recommend physical fitness and proper dieting to help build counter-immunity against COVID-19, especially when immune-compromised individuals are being affected the most. AR/VR has been confirmed to support mobile fitness (Alturki & Gay, 2019).

Pokemon Go mobile app has AR integrated technology meant to support physical fitness. AR can be prevalent in tourism as travelers can make use of AR travel guide apps as an alternative device to substitute travel experience, especially during lockdowns (Adhani, 2012). Rohs and Gfeller (2004) proposed the use of apps built hardware devices configured in a computational platform to aid global tracking of COVID-19 data and control of virus spread.

There are several reasons why AR/VR devices are in high demand in recent times. Kim and Dey (2009) disclosed the use of AR in vehicle windshield displays for elderly people, which support cognitive mapping, and direction to the right destination. Hahn (2012) introduced AR for library service, which can be browsed as physical books that stack optical and facial recognition; implemented for library navigation. Cape Town University in South Africa measured the impact of AR mobile applications on students studying health science. Its impact reveals learning motivation because it grabs attention, offers relevance, boosts confidence and educational satisfaction. The motivational context model of AR learning models, referred to as the ARCS model, aids education attention, relevance, confidence, and satisfaction.

The George Washington Hospital decided to use VR to assess patients with throat and lung damage, enabling them to figure outpatients with COVID-19. The University of California once conducted in-depth research with sensing VR headset and camera, manipulating a three-dimensional model to examine patients with SARS infection. To attain an accurate result, they used R headset, receptor, angiotensin-converting enzymes (ACE-2), and protein cell surface of different tissue types to discover immune challenges in patients.

Hikers at mountain Everest used an hour recreational VR tool to combat travel anxiety. Thus, the google edition of VR app serves tours to international space stations and national museums in Iraq. The google edition of VR can be accomplished using cardboard VR headsets (PowerUpEdu, 2020). The pandemic impacts economic and social lives, leading to a global recession in some developed countries. Covid-19 has re-directed communication, education, and healthcare services. Balsa-Barreiro et al. (2020) declared the pandemic as a genuine reason to commence de-globalization due to lockdown, border restriction, and limited medical supplies. Progressively, it has caused an uninterrupted boom of virtual revolution and strengthens the significance of AR/VR and telemedicine.

Before the lockdown, different countries allowed their doctors and nurses to practice remote Medicare (telemedicine); whereby using video calls, AR/VR to check vital signs and obtain X-ray scans. A perfect example is a South Korean Radiologist who used AR to conduct X-ray scans on patients. The KCDC developed "Now and Here" app to keep the non-infected region from contracting the virus. They also designed the "COBAEK" app to alert people about COVID-19 spread within 100-m distance by sending an alarm to users, which virtually directs people's movement.

EduVenture and Museum Scrabble AR teaching and learning recently ranked the top list of well-documented, pattern and framework analyses that functions as an engaging information exchange program that engages team-spirit, personal learning, motivation and gamified learning (Ferdinand, Müller, Ritschel, & Wechselberger, 2005; Yiannoutsou, Papadimitriou, Komis, & Avouris, 2009). Since coronavirus is prevalently a new virus, medical schools need to offer staff and medical students training on how to use VR to predict COVID-19. VR can significantly be used to treat patients, campaigns, and create disease awareness by showing the simulated outcome of the disease to the public. Medical awareness will help patients to understand the ultimate use of VR and why it's needed for their treatment.

VR can be used to study mental or psychological stress and disturbance in patients with COVID-19. Due to therapeutic procedures, VR creates an atmosphere for motivation and reveals the healing process to patients (Franchi, 2020).

VRs are recorded to be implemented in different contexts of COVID-19 cases, such as physical therapy and pain management. VR application works for both patients in need of mobile or in-hospital treatment. VR has successfully been used to handle physical rehabilitation and cognitive therapy. Physicians confessed positive results and explained how VR distracts patients while reducing psychological traumas experienced by patients undergoing complex operations (Rajkumar, 2020).

4.1 AR/VR and other new technologies for social distancing and controlling the spread of COVID-19

Researchers agreed during lockdown that VR will support social distancing and promote educational transformation that will enable students to rely on virtual classes while adopting luxurious learning styles from home; joining educational and official conferences.

South Korea stepped ahead with CBS (cellular broadcasting service) to exchange emergency messages and communicate effectively with citizens. CBS messages are sent to mobile devices by the Korean telecom company. They configured "Holding Virtual Video Conferencing" tool to enhance virtual reality experiences and restrict in-person meetings that cause the spread of the virus. The Korean government also encouraged the use of Global Virtual Private Network (GVPN) and G-drive to perfect remote working. G-Drive is a cloud storage application that supports file storage and document handling (Kwok & Koh, 2020).

Some elderly people from age 65 normally develop musculoskeletal diseases in the United States, leading to constant visits to the clinic. During the lockdown, such patients might find it difficult to access health services. Clinical innovations for outpatient settings and telehealth integration with VR experience will help patients to receive medical advice and attention without hospital visits (Saad, 2019).

The South Korean Government made a firm determination to flatten the COVID-19 curve by integrating Drive and Walk-Through Testing AR software with an attached mirror testing screening device. AR GPS Drive/Walk-through navigation has location-based integration, which can be different from VR, utilized for digital world visualization (like Oculus Rift and Samsung Gear VR). The Boston Human performance institute developed "Whoop Strap 3.0" app for research collaboration to consider mechanisms that increase COVID-19 incubation (Yannone, 2020).

Estimote Workplace Contact Tracing Application acts as a Bluetooth location device to help trace workplace activities and stay safe in the workplace (Burns, 2020).

Life signal corporations launched biosensor patches for leveraging cardiovascular detection of COVID-19. It can also help monitor early spreads using single-shower-proof check tools placed on the chest area. The single-show-proof check tool will be essential for detecting ECG, respiratory, and temperature rates (NS Medical Devices, 2020).

Spryhealth Company invested in a new digital technology called "Loop Signal" that limits patient's clinic visits while offering a remote signal tracking interface for determining heart rates, respiratory and oximetry pulses (Spry Health, 2020).

Shanghai Public health Clinic center deployed the use of Bluetooth gateway called Cassi Network to monitor Covid-19 spread and minimize human contact (Koh, 2020).

Singaporean Government invested in a contact tracing app with Bluetooth AR called Trace together that determines the history of the infected people and subdue further spread of the virus (Chong, 2020).

GPS (2020) ascertained global positioning satellites by navigating with AR tools to monitor real-time locations of positive patients while simulating the outcome to create awareness.

The ministry of health in Israel developed the Hamagen app to monitor positive cases, measure self-isolation, and minimize COVID-19 spread (Cohen, 2020).

Most voice-based COVID-19 diagnostic apps are displayed on real-time AR/VR to discover throat-breaking sound and multiple parameters that show whether a patient is positive (Das, 2020).

AR/VR can effectively work when installed on the 5G plus network because 5G helps to overcome poor network challenges and enable immersive VR/AR performance (Li & Xia, 2020).

Huawei and Deloitte (2020) discussed 5G plus medical imaging with AR/VR authentication that handles picture archive and communication systems (PACS), required to diagnose and treat viruses. They also mentioned the thermal imaging technology initially deployed for anti-aircraft defense, thus adopting 5G plus IR thermal image monitor for real-time temperature checks with an accuracy rate of 89%. AR/VR notably provides critical medical rehabilitation and training using telemedical procedures that handle robotic support for limbs and exercise rehabilitation.

VR offers the expected streamlined opportunity to review vital signs during teleconsultation. Vital signs screening can be calculated from zero-latency of 1 ms radio latency, whereas VR is projected during surgery to demonstrate the stability of patients' heart rates (Stefano & Kream, 2018).

4.2 Use of AR/VR to support remote education and reduce the spread of COVID-19

Virtual reality creates beyond mainstream experience and offers an immersive real 3D dimensional visual feedback that can engage distracted students. Educational VR has been around since the 1990s but didn't go far because experts considered it expensive and non-realistic. The pandemic brought back the forgotten experiment and established new perspectives utilized as Google cardboard, daydream view, Oculus Rift, HTC Vive, etc.; as high definition VR tools for attaining knowledge-driven experience. AR integration can digitally enhance objects to be viewed as real-world videos using smartphones and cameras. Some AR has language translation and QR scan code integration, which enables the viewing of real images in a virtual environment (Cruz-Niera, 2016).

Delialioglu (2012) discovered that traditional methods of teaching could lead to a lack of educational engagement, causing students to develop truant habits, learning confusion, and dissatisfaction. However, VR allows students to experience learning engagement in interactive and hands-on training. Google expeditions permit virtual trips to Mars for students to learn geography and planets. VR enables students to explore with curiosity; specifically, archeological students that won't travel due to COVID-19 may have the privilege to explore VR-specific tools for exploring age and modern caves and archeological sites (Costa & Melotti, 2012). Since VR gives a strong sense of presence that can replace traditional classrooms, VR can encourage reluctant students to learn with excitement (Aylett & Louchart, 2003). According to experts, VR can perform limitless educational tasks and provides constructivist learning, whereby students can solve real educational problems while collaborating with other students. For instance, astronomy courses are manipulated in VR to offer visual concepts of traditional topics. The same principles of VR learning can be accessible in physics, mathematics, machine-building and learning, and problem-solving (Bailenson et al., 2008).

VR holds other significance for students with post-traumatic stress, disabilities, and social anxiety issues which can learn with the tool (Standen & Brown, 2006). VR tools are available for language and public speech learning which enhances speech and help students with speech phobia (VirtualSpeech, 2016).

Scientists unravel the cost and time wasted using traditional methods of training healthcare professionals and suggest VR tools as a capital-intensive application that reduces durations. AR/VR may enhance a multifaceted physiological system of learning and fill-in the gap between theoretical and practical learning methods, in turn, to provide realistic, complex and dynamic knowledge on how to handle CPRs and other emergency equipment for stabilizing patients with COVID-19 infection (Creutzfeldt, Hedman, & Felländer-Tsai, 2016).

Some VR technologies act as prevention and treatment tools for emergency cases, whereby paramedic trainees are subjected to learn and use AR/VR devices to save lives (Hoge et al., 2004).

Ferriter (2016) tested AR impacts on the textbooks integration approach in Malaysia. Chang and O'Sullivan (2005) reviewed AR mobile learning systems on natural science for fourth-grade students in Taiwan. Akçayır and Akçayır (2017) implored the use of AR interfaces to enhance laboratory science in turkey, by augmenting 4D anatomy testing for undergraduate students. AR/VR seems quite significant in education and holds the future of education due to its hybrid learning environment, scientific experience, real-world phenomena, and visible chemical reactions.

AR can improve thinking skills, increase understating, and address practical learning difficulties. However, challenges such as complex UI (User-Interface) can affect students' ability to use the tool. Technical and design errors, navigation difficulties and multitasking, inflexibility, and fantasy might influence students' ability to learn.

Surgeons are primarily dependent on AR/VR to train and conduct visual techniques before operations. AR offers a 3D dimensional representation of patients' anatomy while improving outcomes and accuracies of post-operative procedures. Surgeons also depend on AR for vein visualization, specifically when injecting blood or taking blood samples of patients with an invisible vein. The vein augmented process, referred to as "AccuVein," helps physicians to take blood samples of infants and aged adults (Julier & Bishop, 2002). The South Korean Government used the "All-Science" VR portal to tackle the impact of COVID-19 on education. The portal has immersive VR video clips, comics, and quizzes. EDISON (Educational Research Integrated through Simulation Net), has a similar VR portal to provide an international classroom for science and technology students. Most educational broadcasting system (EBS) networks integrate video, text, and data; linked to YouTube and IPTV to provide real-time education globally.

4.3 Limitations and health risks of AR/VR

AR/VR can become addictive if users regularly depend on its stereotype HMD and the head tracker graphics can cause brain damage after a long time of use. Researchers explained that constant use of AR/VR could lead to clusters of confusion because the user may find it difficult to distinguish between virtual events and real events. Another issue mentioned is the ultra-realistic experience that forces emotional instability, tense behavior, and an inability for some users with personality disorders to readjust to real-world experience. People who depend on AR/VR daily may experience social isolation. However, data privacy and identity impersonation are some rising issues of AR/VR in recent times (Slater et al., 2020).

Technical and language barriers are mentioned as limiting factors of using AR/VR in telemedicine. People in rural areas with a poor internet connection and audio quality may face struggles with the tool.

Buvik, Bugge, Knutsen, Småbrekke, and Wilsgaard (2016) recommend post-visits to hospitals if symptoms persist and if dissatisfied, patients can report to compliant cloud-based imaging applications. He also explained common conflicts surrounding the use of AR/VR, such as verging muscle movement conflict, accommodation balance issues, visual discomfort and double vision. However, it is important to control eye sensitivity with an eye-tracker and watch out for issues such as flicker calibration and adjustment issues.

5 Conclusion

Studies made numerous representations of computable near-eye displays with AR/VR applications that coordinate simulations, training, visualization, entertainment, games, robotic surgery, education, travel and architectural sightseeing. AR/VR was noted to have penetrated medical schools, especially the recent utilization of it in physiotherapy, psychotherapy, anxiety and stress treatment, and pain management. For the present COVID-19 situation, AR/VR aids X-ray visualization, Marketing campaign to create awareness of the spread and virtual simulation of COVID-19 data. Previously, it worked as a host technology in cloud engineering, conference streaming, near-eye visual acuity, haptic sensor VR camera authentication, and human-computer interface design. Despite the limitations experienced, its implementation enhances multilayered and micro-display of images in 2-3D dimensional formats. This paper discussed different disciplines where AR/VRs are found obtainable. It also discussed its uses in maintaining social distancing, monitoring/controlling COVID-19 virus, and other unique technologies similar to AR/VR that will help mitigate COVID-19 cases.

References

Abdel-Basset, M., Chang, V., & Mohamed, R. (2020). HSMA_WOA : A hybrid novel slime mould algorithm with whale optimization algorithm for tackling the image segmentation problem of chest X-ray images. *Applied Soft Computing*, *95*, 106642. https://doi.org/10.1016/j.asoc.2020.106642.

Abowd, G. D., & Mynatt, E. D. (2000). Charting past, present, and future research in ubiquitous computing. *ACM Transactions on Computer-Human Interaction*, 7(1), 29–58. https://doi.org/10.1145/344949.344988.

Abriata, L. A. (2020). Building blocks for commodity augmented reality-based molecular visualization and modeling in web browsers. *PeerJ Computer Science*, *6*. https://doi.org/10.7717/peerj-cs.260, e260.

Ackerman, E. (2020). *Zipline wants to bring medical drone delivery to U.S. to fight COVID-19—IEEE Spectrum*. IEEE Spectrum: Technology, Engineering, and Science News. https://spectrum.ieee.org/automaton/robotics/drones/zipline-medical-drone-delivery-covid19.

Adhani, N. (2012). A survey of mobile augmented reality applications. In *1st international conference on future trends in computing and communication Technologies*.

Ahmad, F., & Musilek, P. (2006). A keystroke and pointer control input interface for wearable computers. In *Fourth annual IEEE international conference on pervasive computing and communications (PERCOM'06)* (pp. 10–11). https://doi.org/10.1109/PERCOM.2006.2.

Akçayır, M., & Akçayır, G. (2017). Advantages and challenges associated with augmented reality for education : A systematic review of the literature. *Educational Research Review*, *20*, 1–11. https://doi.org/10.1016/j.edurev.2016.11.002.

Alturki, R., & Gay, V. (2019). Augmented and virtual reality in mobile fitness applications: A survey. In F. Khan, M. A. Jan, & M. Alam (Éds.), Applications of intelligent technologies in healthcare (p. 67–75). Springer International Publishing. doi:https://doi.org/10.1007/978-3-319-96139-2_7.

Amirghofran, Z. (2010). Medicinal plants as immunosuppressive agents in traditional Iranian medicine. *Iranian Journal of Immunology: IJI*, 7(2), 65–73. https://pubmed.ncbi.nlm.nih.gov/20574119/.

Aylett, R., & Louchart, S. (2003). Towards a narrative theory of virtual reality. *Virtual Reality*, 7(1), 2–9. https://doi.org/10.1007/s10055-003-0114-9.

Azuma, R., Baillot, Y., Behringer, R., Feiner, S., Julier, S., & Mac Intyre, B. (2001). Recent advances in augmented reality. *IEEE Computer Graphics and Applications*, *21*(6), 34–47. https://doi.org/10.1109/38.963459.

Bailenson, J. N., Yee, N., Blascovich, J., Beall, A. C., Lundblad, N., & Jin, M. (2008). The use of immersive virtual reality in the learning sciences: Digital transformations of teachers, students, and social context. *Journal of the Learning Sciences*, *17*(1), 102–140. https://doi.org/10.1080/10508400701793141. VHIL.

Balsa-Barreiro, J., Vié, A., Morales, A. J., & Cebrián, M. (2020). Deglobalization in a hyper-connected world. *Palgrave Communications*, *6*(1), 1–4. https://doi.org/10.1057/s41599-020-0403-x.

Benford, S., Greenhalgh, C., Reynard, G., Brown, C., & Koleva, B. (1998). Understanding and constructing shared spaces with mixed-reality boundaries. *ACM Transactions on Computer-Human Interaction*, *5*(3), 185–223. https://doi.org/10.1145/292834.292836.

Burns, C. (2020, April 2). *Estimote wearables track workers to curb COVID-19 outbreak*. SlashGear. https://www.slashgear.com/estimote-wearables-track-workers-to-curb-covid-19-outbreak-02615366/.

Businesswire. (2020, June 26). *COVID-19 impact and recovery analysis- global augmented reality (AR) and virtual reality (VR) market 2020–2024 | Increasing demand for AR and VR technology to boost market growth*. Technavio. https://www.businesswire.com/news/home/20200626005010/en/COVID-19-Impact-and-Recovery-Analysis- -Global-Augmented-Reality-AR-and-Virtual-Reality-VR-Market-2020-2024-Increasing-Demand-for-AR-and-VR-Technology-to-Boost-Market-Growth-Technavio.

Buvik, A., Bugge, E., Knutsen, G., Småbrekke, A., & Wilsgaard, T. (2016). Quality of care for remote orthopaedic consultations using telemedicine : A randomised controlled trial. *BMC Health Services Research*, *16*, 483. https://doi.org/10.1186/s12913-016-1717-7.

Carlson, K. M., Curran, L. M., Asner, G. P., Pittman, A. M., Trigg, S. N., & Marion Adeney, J. (2013). Carbon emissions from forest conversion by Kalimantan oil palm plantations. *Nature Climate Change*, *3*(3), 283–287. https://doi.org/10.1038/nclimate1702.

Chandler, S. (2020, July 2). *Meet the companies using VR To treat coronavirus-related stress and anxiety*. Forbes. https://www.forbes.com/sites/simonchandler/2020/07/02/meet-the-companies-using-vr-to-treat-coronavirus-related-stress-and-anxiety/.

Chandola, S. (2020, May 11). *Returning to a new normal with AR & VR*. Geospatial World. https://www.geospatialworld.net/blogs/returning-to-a-new-normal-with-ar-vr/.

Chang, A., & O'Sullivan, C. (2005). Audio-haptic feedback in mobile phones. In *CHI '05 extended abstracts on human factors in computing systems* (pp. 1264–1267). https://doi.org/10.1145/1056808.1056892.

Chehab, M. A., Brinjikji, W., Copelan, A., & Venkatesan, A. M. (2015). Navigational tools for interventional radiology and interventional oncology applications. *Seminars in Interventional Radiology*, *32*(4), 416–427. https://doi.org/10.1055/s-0035-1564705.

Chong, C. (2020, April 1). *About 1 million people have downloaded TraceTogether app, but more need to do so for it to be effective : Lawrence Wong [text]*. The straits times. https://www.straitstimes.com/singapore/about-one-million-people-have-downloaded-the-tracetogether-app-but-more-need-to-do-so-for.

Cohen, T. (2020, April 1). *1.5 million Israelis using voluntary coronavirus monitoring app*. Reuters. https://www.reuters.com/article/us-health-coronavirus-israel-apps-idUSKBN21J5L5.

Costa, N., & Melotti, M. (2012). Digital media in archaeological areas, virtual reality, authenticity and hyper-tourist gaze. *Sociology Mind*, *2*(1), 53–60. https://doi.org/10.4236/sm.2012.21007.

Creutzfeldt, J., Hedman, L., & Felländer-Tsai, L. (2016). Cardiopulmonary resuscitation training by avatars : A qualitative study of medical Students' experiences using a multiplayer virtual world. *JMIR Serious Games*, *4*(2). https://doi.org/10.2196/games.6448.

Cruz-Niera, C. (2016). Beyond fun and games : VR as a tool of the trade. In *Session presented at the virtual reality summit* (p. 12). New York, NY, USA.

Das, M. (2020, March 29). *Role of AI soars in tackling Covid-19 pandemic*. Businessline. https://www.thehindubusinessline.com/info-tech/role-of-ai-soars-in-tackling-covid-19-pandemic/article31197098.ece.

Delialioglu, O. (2012). Student engagement in blended learning environments with lecture-based and problem-based instructional approaches. *Journal of Educational Technology and Society*, *15*(3), 310–322.

Egor, P., Shoaib, S., Sartaaj, W., Michael, O., Lin, C., Mitchell, O.'. L., et al. (2018). Mp26-20 virtual-reality projected renal models with urolithiasis as an educational and preoperative planning tool for nephrolithotomy : A pilot study. *Journal of Urology*, *199*(4S), e345. https://doi.org/10.1016/j.juro.2018.02.878.

Ferdinand, P., Müller, S., Ritschel, T., & Wechselberger, U. (2005). *The eduventure—A new approach of digital game based learning combining virtual and mobile augmented reality games episodes*. https://www.bibsonomy.org/bibtex/2280f0d5d47f32b1a9e8ea5905ca3ba6f/wechselberger.

Ferriter, B. (2016, March 9). Tool review : Google expeditions virtual reality app. In *The Tempered Radical.* https://blog.williamferriter.com/2016/03/09/tool-review-googleexpeditions-virtual-reality-app/.

Ferrucci, L., Perry, J. R. B., Matteini, A., Perola, M., Tanaka, T., Silander, K., et al. (2009). Common variation in the β-carotene 15,15′-monooxygenase 1 gene affects circulating levels of carotenoids : A genome-wide association study. *American Journal of Human Genetics, 84*(2), 123–133. https://doi.org/10.1016/j.ajhg.2008.12.019.

Fjeld, M., & Voegtli, B. M. (2002). Augmented chemistry : An interactive educational workbench. In *Proceedings. International symposium on mixed and augmented reality* (pp. 259–321). https://doi.org/10.1109/ISMAR.2002.1115100.

Franchi, T. (2020). The impact of the Covid-19 pandemic on current anatomy education and future careers : A Student's perspective. *Anatomical Sciences Education, 13*(3), 312–315. https://doi.org/10.1002/ase.1966.

Fuchs, H., State, A., Pisano, E. D., Garrett, W. F., Hirota, G., Livingston, M., Whitton, M. C., & Pizer, S. M. (1996). Towards performing ultrasound-guided needle biopsies from within a head-mounted display. In K. H. Höhne & R. Kikinis (Éds.), Visualization in biomedical computing(p. 591–600). Springer. doi:https://doi.org/10.1007/BFb0047002.

Gotra, A., Sivakumaran, L., Chartrand, G., Vu, K.-N., Vandenbroucke-Menu, F., Kauffmann, C., et al. (2017). Liver segmentation : Indications, techniques and future directions. *Insights Into Imaging, 8*(4), 377–392. https://doi.org/10.1007/s13244-017-0558-1.

GPS. (2020). *GPS overview. GPS.gov.* https://www.gps.gov/systems/gps/.

Greenfield, M. J., Luck, J., Billingsley, M. L., Heyes, R., Smith, O. J., Mosahebi, A., et al. (2018). Demonstration of the effectiveness of augmented reality telesurgery in complex hand reconstruction in Gaza. *Plastic and Reconstructive Surgery. Global Open, 6*(3). https://doi.org/10.1097/GOX.0000000000001708, e1708.

Hahn, J. (2012). Mobile augmented reality applications for library services. *New Library World, 113*(9/10), 429–438. https://doi.org/10.1108/03074801211273902.

Held, R. T., & Hui, T. T. (2011). A guide to stereoscopic 3D displays in medicine. *Academic Radiology, 18*(8), 1035–1048. https://doi.org/10.1016/j.acra.2011.04.005.

Hildrew, D. M. (2020, March 18). *Prioritizing novel approaches to telehealth for all practitioners [text].* American Academy of Otolaryngology-Head and Neck Surgery. https://www.entnet.org/content/prioritizing-novel-approaches-telehealth-all-practitioners.

Hoge, C. W., Castro, C. A., Messer, S. C., McGurk, D., Cotting, D. I., & Koffman, R. L. (2004). Combat duty in Iraq and Afghanistan, mental health problems, and barriers to care. *The New England Journal of Medicine, 351*(1), 13–22. https://doi.org/10.1056/NEJMoa040603.

Hollander, J. E., & Carr, B. G. (2020). Virtually perfect? Telemedicine for Covid-19. *New England Journal of Medicine.* https://doi.org/10.1056/NEJMp2003539.

Huawei, & Deloitte. (2020, March 19). *Huawei and Deloitte publish a whitepaper on combating COVID-19 with 5G.* Channel Post MEA. http://channelpostmea.com/2020/03/19/huawei-deloitte-publish-whitepaper-combating-covid-19-5g/.

Ishii, H., & Ullmer, B. (1997). Tangible bits : Towards seamless interfaces between people, bits and atoms. In *Proceedings of the ACM SIGCHI conference on human factors in computing systems* (pp. 234–241). https://doi.org/10.1145/258549.258715.

IWAR'99. (1999). *Proceedings 2nd IEEE and ACM international workshop on augmented reality (IWAR'99). ii-ii.* https://doi.org/10.1109/IWAR.1999.803800.

Julier, S., & Bishop, G. (2002). Tracking : How hard can it be? *IEEE Computer Graphics and Applications, 22*(6), 22–23. https://doi.org/10.1109/MCG.2002.1046625.

Kato, H., & Billinghurst, M. (1999). Marker tracking and HMD calibration for a video-based augmented reality conferencing system. In *Proceedings 2nd IEEE and ACM international workshop on augmented reality (IWAR'99)* (pp. 85–94). https://doi.org/10.1109/IWAR.1999.803809.

Kim, S., & Dey, A. K. (2009). Simulated augmented reality windshield display as a cognitive mapping aid for elder driver navigation. In *Proceedings of the SIGCHI conference on human factors in computing systems* (pp. 133–142). https://doi.org/10.1145/1518701.1518724.

Koh, D. (2020, April 3). *SPHCC employs IoT tech and wearable sensors to monitor COVID-19 patients.* MobiHealthNews. https://www.mobihealthnews.com/news/apac/sphcc-employs-iot-tech-and-wearable-sensors-monitor-covid-19-patients.

Kwok, A. O. J., & Koh, S. G. M. (2020). COVID-19 and extended reality (XR). *Current Issues in Tourism*, 1–6. https://doi.org/10.1080/13683500.2020.1798896.

Li, Y., & Xia, L. (2020). Coronavirus disease 2019 (COVID-19) : Role of chest CT in diagnosis and management. *AJR. American Journal of Roentgenology*, *214*(6), 1280–1286. https://doi.org/10.2214/AJR.20.22954.

Light, D., & Brown, L. (2020). Dwelling-mobility : A theory of the existential pull between home and away. *Annals of Tourism Research*, *81*, 102880. https://doi.org/10.1016/j.annals.2020.102880.

López-Pérez, M. V., Pérez-López, M. C., Rodríguez-Ariza, L., & Argente-Linares, E. (2013). The influence of the use of technology on student outcomes in a blended learning context. *Educational Technology Research and Development*, *61*(4), 625–638. https://doi.org/10.1007/s11423-013-9303-8.

Lu, W., Wang, L., Zhou, W., & Fu, W. (2019). Augmented reality navigation to assist retrograde peroneal access for the endovascular treatment of critical limb ischemia. *Journal of Vascular Surgery Cases and Innovative Techniques*, *5*(4), 518–520. https://doi.org/10.1016/j.jvscit.2019.06.004.

Microsoft. (2019). *Matériel HoloLens (1ère génération).* https://docs.microsoft.com/fr-fr/hololens/hololens1-hardware.

NS Medical Devices. (2020, April 3). *LifeSignals to roll out biosensor patch for COVID-19 monitoring.* NS Medical Devices. https://www.nsmedicaldevices.com/news/lifesignals-biosensor-patch-covid-19/.

Papa, S. M., Brundin, P., Fung, V. S. C., Kang, U. J., Burn, D. J., Colosimo, C., et al. (2020). Impact of the COVID-19 pandemic on Parkinson's disease and movement disorders. *Movement Disorders: Official Journal of the Movement Disorder Society*, *35*(5), 711–715. https://doi.org/10.1002/mds.28067.

Perkins, C. (2020). *2020 augmented & virtual reality survey results.* Perkins Coie. https://www.perkinscoie.com/en/ar-vr-survey-results/2020-augmented-and-virtual-reality-survey-results.html.

PowerUpEdu. (2020). *Career pathways and COVID19.* https://powerupedu.com/wp-content/uploads/2020/05/Career-Pathways-and-COVID19.doc-Rev-PUEDU.pdf. (Accessed 1 April 2021).

Rajkumar, R. P. (2020). COVID-19 and mental health : A review of the existing literature. *Asian Journal of Psychiatry*, *52*, 102066. https://doi.org/10.1016/j.ajp.2020.102066.

Rohs, M., & Gfeller, B. (2004). Using camera-equipped mobile phones for interacting with real-world objects. In *Advances in pervasive computing* (pp. 265–271).

Saad, L. (2019, December 9). *More Americans delaying medical treatment due to cost.* Gallup.Com. https://news.gallup.com/poll/269138/americans-delaying-medical-treatment-due-cost.aspx.

Sirakaya, M., & Sirakaya, D. A. (2018). Trends in educational augmented reality studies : A systematic review. *Malaysian Online Journal of Educational Technology*, *6*(2), 60–74.

Slater, M., Gonzalez-Liencres, C., Haggard, P., Vinkers, C., Gregory-Clarke, R., Jelley, S., et al. (2020). The ethics of realism in virtual and augmented reality. *Virtual Reality*, *1*-(1). https://doi.org/10.3389/frvir.2020.00001.

Spry Health. (2020, March 20). *Spry health launches new service, loop signal, to assist in surge of COVID-19 cases by reducing avoidable hospital visits and improving patient monitoring at home.* https://www.businesswire.com/news/home/20200320005135/en/Spry-Health-Launches-New-Service-Loop-Signal-to-Assist-in-Surge-of-COVID-19-Cases-by-Reducing-Avoidable-Hospital-Visits-and-Improving-Patient-Monitoring-at-Home.

Standen, P. J., & Brown, D. J. (2006). Virtual reality and its role in removing the barriers that turn cognitive impairments into intellectual disability. *Virtual Reality, 10*(3), 241–252. https://doi.org/10.1007/s10055-006-0042-6.

Stefano, G. B., & Kream, R. M. (2018). The micro-hospital : 5G telemedicine-based care. *Medical Science Monitor Basic Research, 24*, 103–104. https://doi.org/10.12659/MSMBR.911436.

Sutherland, I. E. (1965). The ultimate display. In *Proceedings of the IFIP congress* (pp. 506–508).

Tang, A., Owen, C., Biocca, F., & Mou, W. (2002). Experimental evaluation of augmented reality in object assembly task. In *Proceedings. International symposium on mixed and augmented reality* (pp. 265–266). https://doi.org/10.1109/ISMAR.2002.1115105.

Unity Technologies. (2005). *Unity real-time development platform | 3D, 2D VR & AR engine.* https://unity.com/.

Vávra, P., Roman, J., Zonča, P., Ihnát, P., Němec, M., Kumar, J., et al. (2017). Recent development of augmented reality in surgery : A review. *Journal of Healthcare Engineering, 2017*, 4574172. https://doi.org/10.1155/2017/4574172.

VirtualSpeech. (2016). *VirtualSpeech—Soft skills training with vr and simulations.* https://virtualspeech.com.

Wong, H. Y. F., Lam, H. Y. S., Fong, A. H.-T., Leung, S. T., Chin, T. W.-Y., Lo, C. S. Y., et al. (2020). Frequency and distribution of chest radiographic findings in patients positive for COVID-19. *Radiology, 296*(2), E72–E78. https://doi.org/10.1148/radiol.2020201160.

Xu, M., & Hua, H. (2020). Co-axial depth sensor with an extended depth range for AR/VR applications. *Virtual Reality & Intelligent Hardware, 2*(1), 1–11. https://doi.org/10.1016/j.vrih.2019.10.004.

Yannone, T. (2020, March 4). *Could fitness wearables help detect early signs of COVID-19?.* https://www.bostonmagazine.com/health/2020/04/03/fitness-wearables-coronavirus/.

Yiannoutsou, N., Papadimitriou, I., Komis, V., & Avouris, N. (2009). "Playing with" museum exhibits: Designing educational games mediated by mobile technology. In *Proceedings of the 8th international conference on interaction design and children* (pp. 230–233). https://doi.org/10.1145/1551788.1551837.

CHAPTER 5

Using interpretable machine learning identify factors contributing to COVID-19 cases in the United States

Ashish Garg
Walmart Global Tech, Bentonville, AR, United States

1 Introduction

1.1 Background of the study

In December 2019, in Wuhan, Hubei, China, several people who visited the wet seafood market were found infected with an unknown virus and the pathogen was similar to coronavirus (CoV). This unknown virus was named 2019 novel coronavirus or 2019-nCoV, which was later renamed by World Health Organization (WHO) as COVID-19 on 11th February 2020 (Coronavirus (COVID-19) events as they happen, n.d.; Sun et al., 2020). Fig. 1—COVID-19 Expansion shows the elaboration of COVID-19, which states what it stands for.

International Committee on Taxonomy of Viruses-Coronaviridae Study Group (ICTV-CSG) of Viruses named COVID-19 as SARS-CoV-2 based on naming nomenclature (Gorbalenya et al., 2020). Fig. 2—Coronavirus naming during the three major outbreaks shows the evolution of the names of different respiratory pandemics.

With the rate at which this virus was spreading, on 30th January 2020, WHO declared this as "Public Health Emergency of International Concern" and later on 11th March 2020 declared this as "pandemic" (Coronavirus (COVID-19) events as they happen, n.d.).

Coronaviruses are genetically diverse viruses found in animals and humans, which cause intestinal and respiratory infections. These were first identified in, 2002–03 in Guangdong Province, which was named as Severe Acute Respiratory Syndrome (SARS). This virus spread over 26 countries, infecting 8096 people with a fatality rate of 10% (WHO | Summary of probable SARS cases with onset of illness from 1st November 2002 to 31st July

Novel AI and Data Science Advancements for Sustainability in the Era of COVID-19
https://doi.org/10.1016/B978-0-323-90054-6.00008-8

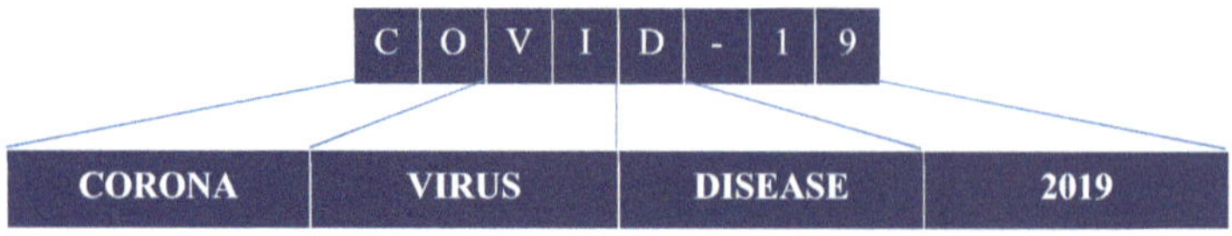

Fig. 1 COVID-19 expansion (Coronavirus (COVID-19) frequently asked questions | CDC, n.d.).

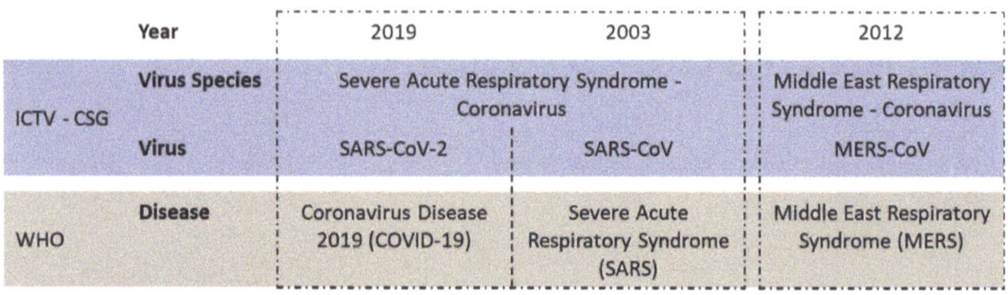

Fig. 2 Coronavirus naming during the three major outbreaks (Gorbalenya et al., 2020).

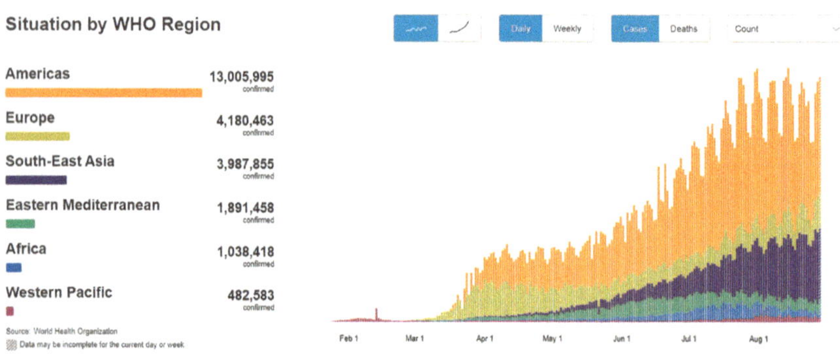

Fig. 3 World dashboard for COVID-19 (WHO coronavirus disease (COVID-19) dashboard | WHO Coronavirus Disease (COVID-19) Dashboard, n.d.).

2003, 2015). It was later found that SARS originated from bats and was transmitted to humans through Himalayan palm civets or raccoon dogs. Another type of virus that originated from an animal is Middle Eastern Respiratory Syndrome (MERS). This virus infected 2494 people with a fatality rate of 35% (WHO | Middle East respiratory syndrome coronavirus (MERS-CoV), 2020).

As of 24:00 PST on 29th August 2020, we have a total of 24,587,513 cases worldwide with 833,556 deaths with the maximum number of confirmed cases in the United States. Fig. 3—World dashboard for COVID-19 shows the spread of cases throughout the world as of 29th August 2020.

Right now we are months away from vaccines and need to use Non-pharmaceutical Interventions (NPIs) to curb the spread of the virus till we have vaccines available (Coronavirus Vaccine Tracker—The New York Times, n.d.; Nonpharmaceutical Interventions (NPIs) | CDC, n.d.). With the option to impose multiple NPIs it becomes difficult to choose which NPI/s will suppress new confirmed cases and sustain the same. Imposing a wrong NPI/s won't impact new confirmed cases but moreover will adversely impact the economy and further create panic in people. To solve this problem, we need to understand mobility patterns of people against new confirmed cases, and this will help in drawing the inference of which mobility pattern spikes up the number of cases based on which an educated decision of imposing NPIs can be considered. A well decided imposed NPI/s will suppress current confirmed cases and sustain the same until it is in effect. This will further help in controlling peak healthcare demand and managing resources effectively.

1.2 Problem statement

COVID-19 is the worst pandemic that the world has experienced in the recent decade. It has claimed more than 800 thousand lives worldwide, with a close 25 million confirmed cases as of 29th August 2020 (WHO coronavirus disease (COVID-19) dashboard | WHO Coronavirus Disease (COVID-19) Dashboard, n.d.). It is on a sharp rise and is becoming more severe with every passing day. The United States is the worst hit by this pandemic, having more than half of total confirmed cases worldwide.

As of 29th August 2020, we stand nearly the past 5 months since COVID-19 was termed as a pandemic by the World Health Organization (WHO) with no identified treatment or vaccine available for the same (Coronavirus (COVID-19) events as they happen, n.d.). Several health organizations are rushing towards the development of vaccines and are trying to expedite the development process, but still, we stand months away from the same (Coronavirus Vaccine Tracker—The New York Times, n.d.). The reason being several complications, regulations, and testing around the vaccine. A normal vaccine typically takes 18 months from inception to first delivery, even though with the expedition of process, time can be slightly reduced but not drastically. Once the vaccine is identified, mass production and availability for the public will again have an added timeframe. COVID-19 spreads from person to person through aerosols or droplets, which can be transmitted through contact routes or respiratory droplets.

Hence, until the vaccine gets available for the public, this pandemic needs to be controlled using other ways known as Non-pharmaceutical Interventions (NPIs).

Non-pharmaceutical Interventions (NPIs) are behavioral restrictions that are adhered to by an individual and as a society to curb the spread of the virus (Nonpharmaceutical Interventions (NPIs) | CDC, n.d.). Several countries have imposed such NPIs in the form of Social Distancing, closing down public places, schools, colleges, workplaces, etc. Imposing NPIs shows a positive effect on curbing the spread of the virus but harms the economy drastically. Hence, NPIs should be implemented to make maximum impact in curbing virus and make the least impact on the economy. Understanding of public mobility patterns with population density and confirmed cases will help in the understanding of the contribution of a place in the spread of COVID-19. This will further help in deciding which particular NPIs should be imposed, which will reduce interaction in that place leading to the curb of the spread of the virus.

Based on this thought, this study focuses on finding NPIs that should be imposed in the United States to curb the spread of COVID-19 in the absence of vaccines based on mobility patterns of public, population data, confirmed cases, and intensity of spread of COVID-19 in a particular region.

1.3 Aim and objectives

The primary aim of this study is to help make the data-backed decision on which NPIs should be imposed to curb the spread of COVID-19, which can be achieved by understanding the impact of the global mobility pattern of people along with population and population density in a region against new confirmed COVID-19 cases observed.

Based on this, the objectives can be

- To evaluate pattern and relationship between mobility factors, population, population density, and new active cases.
- To evaluate the performance of different machine learning models in predicting the number of active cases on a particular day.
- To deduce the interpretability of the model in predicting new active cases
- To identify features that have a high impact on new active cases.
- To identify the statistical model which is efficient in predicting new confirmed cases.

1.4 Research questions

Following research questions have been identified for this study.

- How does the different activity of people in places such as retail stores, parks, shopping center, residential area, etc. impact new confirmed COVID-19 cases.
- Is there any correlation of new confirmed COVID-19 cases with population and population density? COVID-19 has a high R_0, which means a high reproduction rate, so does places with a high population or population density are more prone to have a high number of COVID-19 cases.
- Do existing confirmed cases affect new confirmed COVID-19 cases. It has been found COVID-19 is spread from humans to humans, but we want to find out statistically there is any relationship between new confirmed cases with existing confirmed cases.
- Which features impact most new confirmed COVID-19 cases? This will help in determining which NPIs should be imposed to curb the spread of COVID-19.
- Which statistical model best predicts new confirmed COVID-19 cases using mobility features, population, and population density.

1.5 Significance of the study

- The research will explore how population intensity in different public places impacts new confirmed COVID-19 cases. These findings will help in deciding on which Nonpharmaceutical Interventions (NPIs) should be imposed to curb the spread of COVID-19.
- This research will also focus on understanding the impact of population and population density on new confirmed COVID-19 cases. This will help in understanding how prone a state is to have a severe COVID-19 outbreak.
- Another part of the study is to predict the number of new cases on a given day, and this will help in better planning and management of health services. We will also compare different models and their prediction efficiency to bring a contrast between models and identify the best model for this research.

1.6 Structure of the study

In this research, we follow a sequential approach. We start with "WHY" do we need to perform this research aiming at finding factors contributing to the spread of COVID-19. It starts with the background of the study and dives into the problem statement, followed by the aim, objectives, and research questions.

Next in Section 2, we dive into reviewing existing literature that considers similar interests with COVID-19 and understands the point of view of other researchers. In Sections 2.1 and 2.2, we discuss different attributes of COVID-19 that help in understanding it better, such as "Basic Reproduction Number," which explains how the rate at which infection is spreading, and the "Incubation Period" when the person infected starts showing symptoms. Post that in Section 2.3, we dive into where we stand with treatments and vaccines for the same, also understand the vaccine lifecycle. With this in Section 2.4, we also dive into other measures that can be leveraged in the absence of a vaccine to curb the spread, known as Non-pharmaceutical Interventions (NPIs). Further in Section 2.5, we discuss the strategy for implementing NPIs and Contact Tracing by different countries and evaluate their effect on COVID-19 as well as other socio-economic aspects.

Section 3 is about "HOW" we are going to conduct our research. We discuss different statistical models for this research along with evaluation metrics and steps to obtain model interpretability to understand the magnitude of the impact of features on new confirmed COVID-19 cases.

In Section 4 we evaluate experiments and results, defining the best model and further understanding from model interpretability. Finally, in Section 5 conclusion and future work are explained.

2 Related work

COVID-19, which has been declared a pandemic, is still on a streak of spread, with active cases multiplying every passing day. In this section, we will focus on elaborating on the dynamics of COVID-19, the current state of vaccine development, measures that can help curb spread in the absence of a vaccine, and the type of data that we are considering performing this research.

Fig. 4—Daily new confirmed COVID-19 cases above show the trend of new confirmed cases in different countries; currently, testing is implemented in a full-blown method, but still, it is limited. The number of confirmed cases is directly proportional to testing and actual cases may be more than confirmed cases (Moustakas, 2020). We can see in the given comparison the United States tops the chart with approximately the same high number

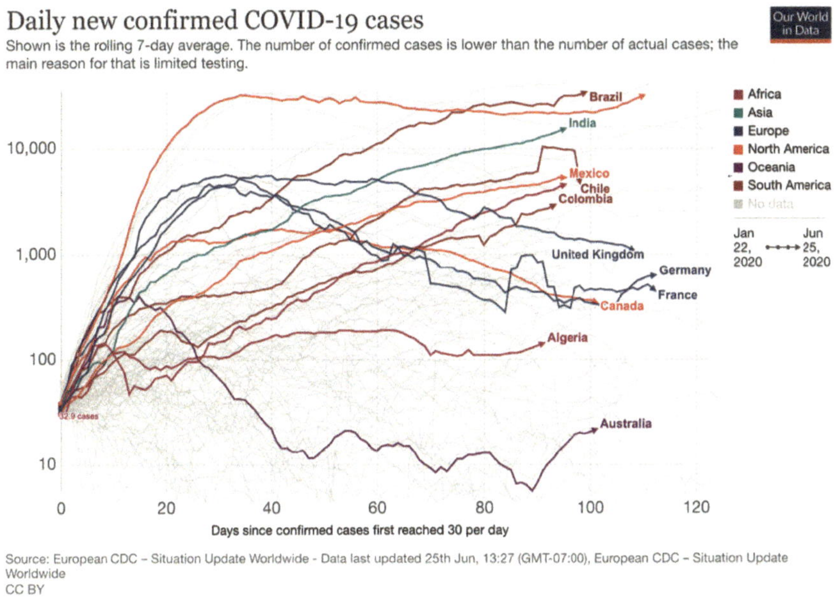

Fig. 4 Daily new confirmed COVID-19 cases (Coronavirus Pandemic Data Explorer—Our World in Data, n.d.).

of confirmed cases from the past 60 days. Community spread has been confirmed in the United States, with New York having the worst outbreak.

COVID-19 virus can spread through respiratory droplets, which can pass from one person to another through verbal interaction, cough, or sneeze when the distance between the 2 is less than 1 m. Another way of transmission is contaminated surface to hands and then to nose, mouth, or eyes passing virus (Ong et al., 2020). Most of the countries have implemented various measures to curb the spread of the virus in the absence of vaccines which all aim towards reducing contact between people resulting in reducing the spread of the virus. These measures have been implemented in the form of Social Distancing, Lockdowns, Ban on public/social gatherings, workplace, school, college closure. These measures have positively impacted reducing the new number of confirmed cases, but a spike is always observed whenever these measures are relaxed.

2.1 Basic reproduction number (R_0)

COVID-19 confirmed cases are increasing exponentially throughout the world, which is possible due to the high Basic Reproduction Rate. Basic Reproduction Number (R_0 pronounced as R naught) is used to measure the transmission dynamics of infections, viruses, and parasites. It is an epidemiologic metric with a numerical value, the higher the value more the virus has an affinity in spreading (Delamater, Street, Leslie, Yang, & Jacobsen, 2019; Dietz, 1993; van den Driessche & Watmough, 2008). Researches following Table 1 R_0 estimates from different published papers demonstrate R_0 calculated by different researchers using different statistical methods. Interestingly researchers have defined the range of R_0 to be in 1.5–6.49 with a mean at 3.28, which is way higher than the R_0 range provided by WHO of 1.4–2.5 with a mean at 1.95.

Table 1 R_0 estimates from different published papers (Liu, Gayle, Wilder-Smith, & Rocklöv, 2020).

Study	Location	Study date	Methods	R_0 estimates
Wu, Leung, and Leung (2020)	Wuhan	31st December 2019–28th January 2020	Stochastic Markov Chain Monte Carlo methods (MCMC)	2.68
Shen, Peng, Xiao, and Zhang (2020)	Hubei province	12th–22nd January 2020	Mathematical model	6.49
Liu et al. (2020)	China and overseas	23rd January 2020	Statistical exponential Growth using Poisson regression	2.90
Liu, Hu, et al. (2020)	China and overseas	23rd January 2020	Statistical maximum likelihood estimation	2.92
Read, Bridgen, Cummings, Ho, and Jewell (2020)	China	1st–22nd January 2020	Mathematical transmission model	3.11
Majumder & Mandl, 2020	Wuhan	8th December 2019–26th January 2020	Mathematical Incidence Decay and Exponential Adjustment (IDEA) model	2.0–3.1 ~2.55
WHO	China	18th January 2020	–	1.4–2.5 ~1.95
Cao et al. (2020)	China	23rd January 2020	The mathematical model including compartments Susceptible-Exposed-Infectious- Recovered-Death-Cumulative (SEIRDC)	4.08
Zhao et al. (2020)	China	10th–24th January 2020	Statistical exponential growth model corresponding to an eightfold increase in the reporting rate	2.24
Zhao et al. (2020)	China	10th–24th January 2020	Statistical exponential growth model corresponding to a twofold increase in the reporting rate	3.58
Imai et al. (2020)	Wuhan	18th January 2020	Mathematical model, computational modeling of potential epidemic trajectories	1.5–3.5 ~2.5
Riou and Althaus (2020)	China and overseas	18th January 2020	Stochastic simulations of early outbreak trajectories	2.2
Tang et al. (2020)	China	22nd January 2020	Mathematical SEIR-type epidemiological model	6.47
Li et al. (2020)	China	22nd January 2020	The statistical exponential growth model	2.2
Average R_0				3.28

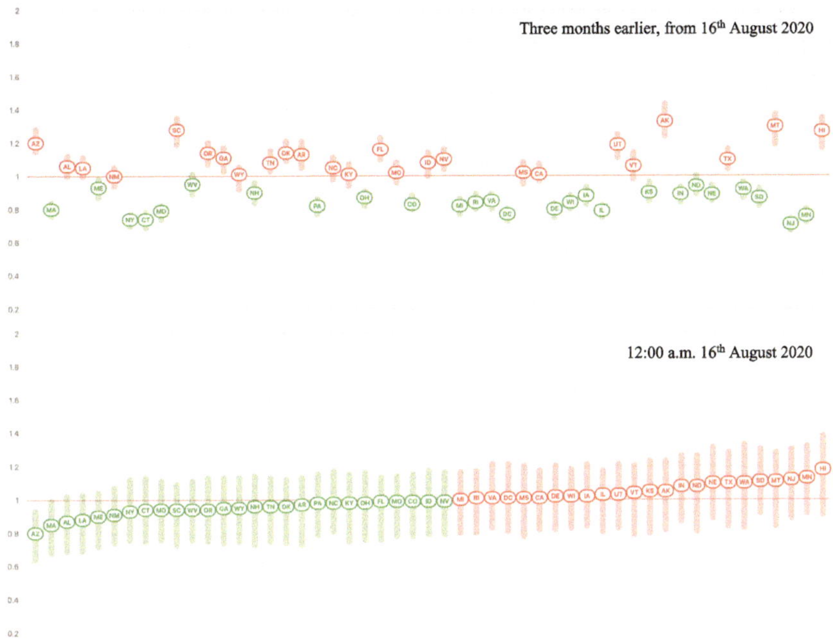

Fig. 5 R_t for different US states (Rt: Effective Reproduction Number, n.d.).

R_0 value of 3.28 interprets that number of infected cases to double within 2 days. These studies consider the early period of December 2019 to January 2020, focusing mainly on cases in China (Sanche et al., 2020).

Real-time R_0 is defined by R_t and the following figure shows R_t for different US states. It is calculated on the go based on current reported new active cases and death numbers. The following figure shows a comparison of R_t for different US states at an interval of 3 months which is May 2020 vs August 2020. We see R_t has dropped for few states while it has increased for a few, this can be attributed to the implementation and relaxation of NPIs in these states. Fig. 5—R_t for different US states shows contrast in R_t for US states with a difference of 3 months. We can see the number of states which had relatively lower R_t had bumped up, while states with higher R_t have stabilized.

2.2 Incubation period and symptoms

Another aspect to look at when considering the spread of COVID-19 is the incubation period. The incubation period is the time between exposure to infection and symptom onset (Lauer et al., 2020). The estimated median incubation period for COVID-19 is 5.1 days with a confidence interval of 95% in the range of 4.5–5.8 days. The majority of cases are 97.5% of cases will show symptoms within 11.5 days. Only 1% of cases will show symptom onset after 14 days. This research is based on confirmed COVID-19 cases reported in the period from 4th January 2020 to 24th February 2020. These results are also confirmed by the Centers for Disease Control and Prevention (CDC) (Management of patients with confirmed 2019-nCoV | CDC, n.d.) and also confirmed that the incubation period is extending to 14 days. From this, we can infer that a person identified as a confirmed COVID-19 patient would have contacted with infection in the last 14 days.

Based on Huang et al. (2020), Chen et al. (2020), and Pan et al. (2020), symptoms are shown by a person contracted with COVID-19 infection differ from person to person, but most people will show the following symptoms, majority reporting fever, cough, fatigue, and anorexia refer Fig. 6—Symptoms shown by COVID-19 cases. Apart from the given symptoms, few people have reported signs and symptoms with lower respiratory tract and gastrointestinal such as diarrhea and nausea before showing up with fever.

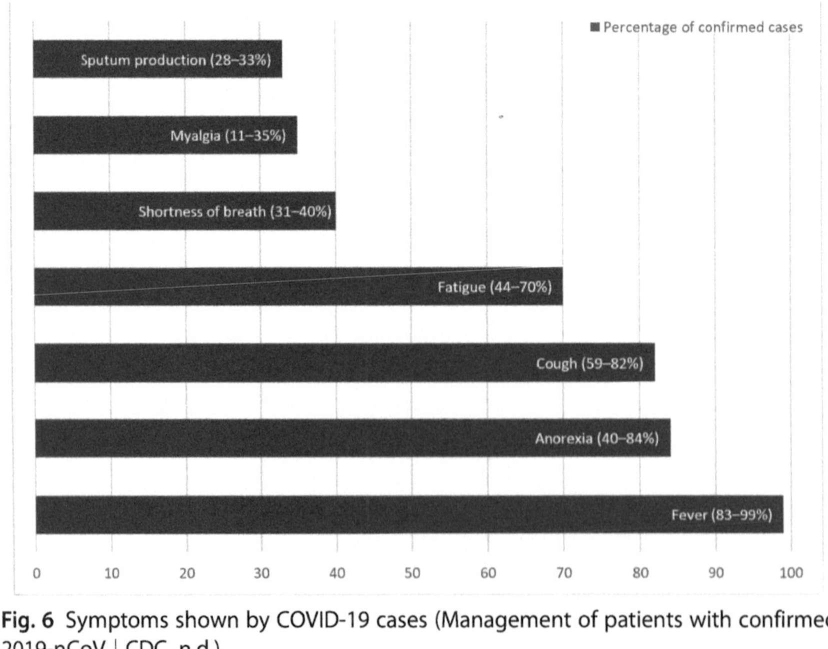

Fig. 6 Symptoms shown by COVID-19 cases (Management of patients with confirmed 2019-nCoV | CDC, n.d.).

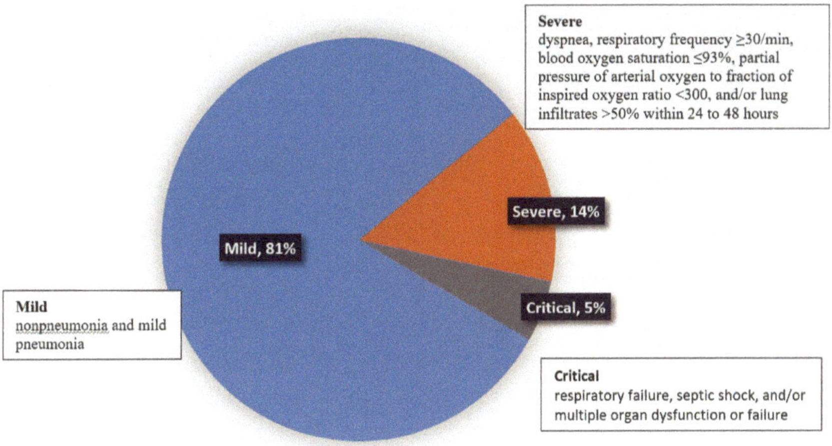

Fig. 7 COVID-19 illness severity.

In a study on 44,672 confirmed cases from the Chinese Center for Disease Control and Prevention performed by (Wu & McGoogan, 2020), the severity of COVID-19 is defined in three categories mild, severe, and critical. Fig. 7—COVID-19 Illness Severity defines a split of confirmed COVID-19 cases among categories mild, severe, and critical with defined symptoms of each category.

2.3 Vaccine trials and availability

With the alarming rate at which COVID-19 is spreading, the World Health Organization (WHO) on 11th March 2020 declared COVID-19 as pandemic (WHO Timeline—COVID-19, n.d.). This also created tremendous pressure worldwide to get a vaccine for the same. Vaccines typically have a timeline of 18 months from the day of inception to be available for public use and have five testing process steps (Coronavirus Vaccine Tracker—The New York Times, n.d.). Fig. 8—Vaccine Testing Process shows different steps involved in vaccine development.

Different agencies throughout the world are rushing in developing vaccines at an increased pace with either combining Phase I and Phase II testing together or approving billions of dollars of funding before it is established that vaccine is efficient. Currently, there are over 140 vaccines in the development stage, with three in Phase III Efficacy Trials, refer to Fig. 9—Vaccine Development Status.

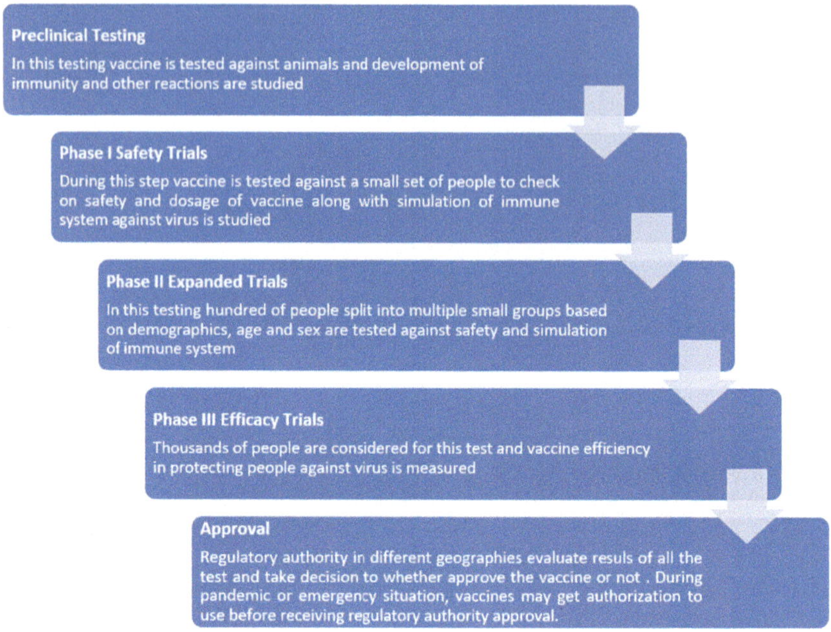

Fig. 8 Vaccine testing process.

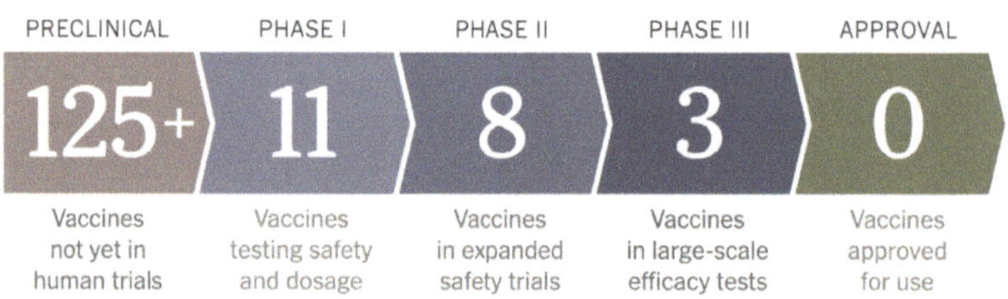

Fig. 9 Vaccine development status (Coronavirus Vaccine Tracker—The New York Times, n.d.).

2.4 Non-pharmaceutical interventions (NPIs)

Even with the current rush on developing vaccines still, there is a good amount of time when the first vaccine will be available and till then other measures need to be put in place to curb the spread of this pandemic. The measures which help curb the spread of virus or infection apart from vaccine or medicine are known as Non-pharmaceutical Interventions (NPIs), which are also known as community mitigation strategies (Nonpharmaceutical Interventions (NPIs) | CDC, n.d.). NPIs define measures are personal, community, and environmental levels to curb the spread

Table 2 NPIs at different levels.

Personal NPI	Community NPI	Environmental NPI
• Staying home when sick • Avoid contact with sick people • Cover coughs and sneezes • Washing hands • Cleaning/ disinfecting frequently touched object or surface	• Maintain social distancing • Temporary closure of facilities which facilitate people are gathering such as schools, colleges, childcare center, sporting events, concerts, movie theaters, etc. • Temporary closure of the workplace and encourage work from home wherever possible	• Routine cleaning or disinfecting of frequently touched objects or surfaces

of the virus, refer to Table 2 NPIs at different levels. Studies from the 1918 influenza pandemic in the United States show that the implementation of Non-pharmaceutical Interventions led to a reduction of mortality rate (Ferguson et al., 2006).

Applying one intervention in a silo does not create a lasting impact, a combination of multiple interventions is required to create a substantial impact (Ferguson et al., 2020). A combination of interventions can be applied using two broad strategies.

1. *Mitigation*—this involves reducing the pace of the spread of pandemics resulting in reduced peak healthcare demand.
2. *Suppression*—this involves reducing new active case numbers and sustaining that going forward.

Comparing the two strategies, "Mitigation" will help in reducing peak healthcare demand. It will just slow down the spread of pandemics but won't impact the number of cases. On the other hand, "Suppression," which requires social distancing of the entire population, quarantine of cases and suspected cases, ban om community, and social gathering along with the closure of schools, universities, and workplaces, will reduce the new case numbers. For suppression to be successful, these interventions should be imposed till the vaccine is available else, whenever these are relaxed a spike in the number of cases will be observed. Since the vaccine development cycle is

18 months, imposing these interventions for such a long duration will come at social and economic costs. In place of implementing interventions for a long period, another way is to implement a combination of interventions in rotation for a small burst of the period, say 3 months, the author has researched implementing such rotation based small duration combination of interventions will result in reducing deaths by half and peak healthcare demand by two-thirds.

2.5 Disruptive technologies

Though we have identified NPI as an interim solution to curb the spread of COVID-19, data is still needed to support where NPIs need to be implemented. Required data is primarily to aid reasoning on whether the target population is likely to have COVID-19, this can be determined based on their social interaction or medical stats. Social interaction provides insights whether they have ever encountered anyone who is carrying COVID-19 or has been in contact with one, whereas medical stats will help in determining how to treat in case of identified positive for COVID-19. Listed below are identified technologies that can effectively implement NPIs (Abdel-Basset, Chang, & Nabeeh, 2021).

2.5.1 Artificial intelligence (AI) and machine learning (ML)

AI and ML are some of the most advanced technologies that are known to us. In these technologies, the machine learns based on existing observation and applies the same to the unknown to derive insights or inferences. These can be used in collaboration with robots for contactless, quick assessment, which is accurate and time-saving. Based on behavioral data, it can also segment users to the susceptibility of getting infected with COVID-19.

2.5.2 Industry 4.0 and internet of medical things (IoMT)

Internet of Things (IoT) is the basis of Industry 4.0 (Abdel-Basset, Nabeeh, El-Ghareeb, & Aboelfetouh, 2020). This brings the advent of medical cyber-physical systems to improve the quality and efficiency of medical services. 360-degree monitoring of the patient through multiple biosensors and applications helps in gaining meaningful insights in a time-efficient manner (Alloghani et al., 2018). This can also be leveraged for the early identification of COVID-19 and treatment of the same in novice state.

2.5.3 Virtual reality (VR), drones and autonomous robots

VR is another advanced technology that simulates a virtual environment, and this helps in creating refreshed, desired experience for the user. This helps psychologically in the current stay-at-home orders (Manto et al., 2020). Drones and autonomous robots are emerging with a vast variety of use cases in healthcare, one of the use cases is contactless high precision telesurgeries (Gupta, Shukla, & Tanwar, 2020; Howe & Matsuoka, 1999). Drones are also used to serve in remote areas for quick delivery of supplies and medical aid. 24×7 Patient observation and surveillance is another key benefit of drones in healthcare.

2.6 Contact tracing

Contact tracing is another important aspect to be taken into account to curb the spread of COVID-19. Any confirmed case is an active carrier of the virus since contact with the virus. Hence it becomes important to trace all the people with whom the confirmed case has been contacted in the last 14 days since the maximum incubation period is deemed 14 days for COVID-19 (Hellewell et al., 2020). For contact tracing and isolation between cases to be effective, the time lag between symptom onset and case isolation should be minimum, along with isolation of people who came in contact with a person showing symptoms. In the case of COVID-19 number of cases and contacts to be traced is huge and also, the availability of testing is not good. So, the feasibility of contact tracing and isolation of cases becomes questionable. In a study that was carried out in Taiwan, it was found that contact tracing and isolating symptomatic patients will only be effective if implemented together with other NPIs such as social distancing (Cheng et al., 2020).

3 Proposed approach

3.1 Dataset description

In this research, we are going to use four datasets providing information about confirmed cases, mobility factors, population, and daily reproduction numbers for each state. For this research, we are going to consider the date range from 8th March 2020 to 30th September 2020 for the country US at the state level. We are considering the given timeframe of data as we have harmonized data available for all US states during this period.

Table 3 Confirmed COVID-19 Cases Dataset Overview lists details around the "Confirmed COVID-19 Cases" dataset.

Table 4 COVID-19 Community Mobility Report Overview lists details around the "COVID-19 Community Mobility Report" dataset.

Table 5 US population—States Overview lists details around the "US population-States" dataset.

Table 3 Confirmed COVID-19 cases dataset overview.

Confirmed COVID-19 cases		
Description	Contains data about confirmed cases daily for different countries. For the United States we have data available at the state and county level. Dataset is in pivot format with dates as columns.	
Key features	• Date • Country	• State • Confirmed cases

Table 4 COVID-19 community mobility report overview.

COVID-19 community mobility report	
Description	Contains data about confirmed cases daily for different countries. For the United States we have data available at the state and county level
Key features	• Date • Country • State • Retail_and_recreation_percent_change_from_baseline • Grocery_and_pharmacy_percent_change_from_baseline • Parks_percent_change_from_baseline • Transit_stations_percent_change_from_baseline • Workplaces_percent_change_from_baseline • Residential_percent_change_from_baseline

Table 5 US population—States overview.

US population—States	
Description	Contains percentage change in mobility factors from baseline for different countries. For the United States, data is available at the state level
Key features	• State • Population • Population density

Table 6 Daily R_0 value for US states lists details around the "Daily R_0 value—US states" dataset.

All of these datasets are available in the public domain and are free to use. All datasets except the population dataset get updated daily.

3.1.1 Smoothing weekend effect

The "Confirmed COVID-19 cases" dataset captures daily new confirmed cases of COVID-19. We notice a pattern where the number of confirmed cases reduces drastically on weekends (Saturday and Sunday) and gets a spike on Monday. This is due to limited testing on weekends and cascading weekend testing to Monday. To harmonize the number of new confirmed cases over the week to eradicate such behavior, we will average out cases from the past 2 days to the future 2 days as shown in Fig. 10—Data transformation of Confirmed COVID-19 cases. Two days past and future are considered to level out any effect of the weekend, which is Saturday and Sunday.

Table 6 Daily R_0 value for US states.

Daily R_0 value—US states	
Description	Contains R_0 values for each state daily
Key features	• Date • Region (State) • Mean R_0

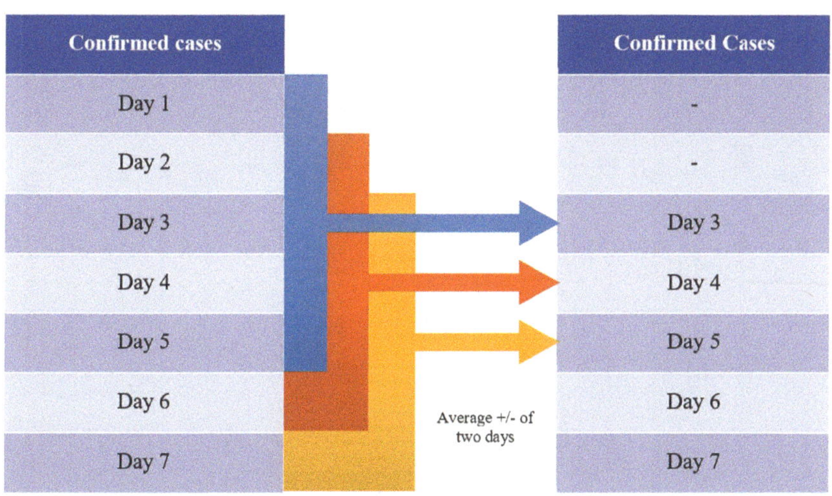

Fig. 10 Data transformation of confirmed COVID-19 cases.

3.2 Feature engineering

During our literature review, we found out that the incubation period COVID-19 virus extends up to 14 days (Management of patients with confirmed 2019-nCoV | CDC, n.d.) from which we can infer a person identified as COVID-19 infected today would have contacted with the virus in past 14 days. Hence, we can consider moving the sum or average of features in the last 14 days as new features to predict confirmed cases today.

3.3 Correlation: Confirmed_cases vs other features

Above Fig. 11—Correlation Confirmed_Cases vs other features shows the correlation of Confirmed_Cases with other features. We notice a relatively good correlation with Cases_past_14_days_mean, Cases_past_14_days_sum, and population which attributes to the fact that COVID-19 spreads from person to person, so an infected person is a carrier of the virus. Hence existing infected cases are directly proportional to new confirmed cases. One of the interesting observations is it is negatively correlated with population density (pop_density) which infers population density does not have any impact on new confirmed cases.

In terms of correlation with mobility features, staying at home is helping curb new confirmed cases, while mobility in retail and recreation areas and, workplaces are helping COVID-19 confirmed cases to bump.

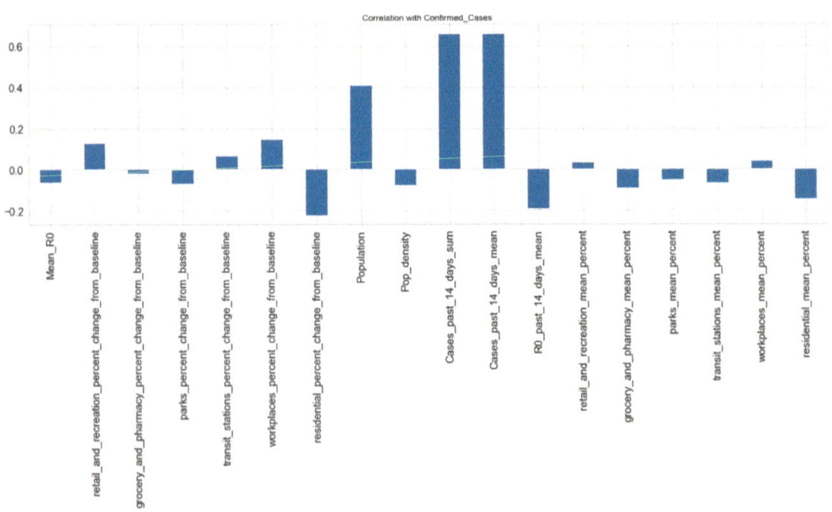

Fig. 11 Correlation confirmed_cases vs other features.

Basic Reproduction Rate (R_0) seems to have a slight negative correlation implying to be an independent feature and does not help in defining new confirmed COVID-19 cases.

4 Experiment and results

4.1 Model building

This is a regression problem for which the following models are considered for evaluation.

- Linear Regression
- Decision Tree
- Random Forest
- Gradient Boosting
- Light Gradient Boosting
- Extreme Gradient Boosting
- AdaBoost

Metrics used for evaluation are R2 and RMSE, these regressor models performed as defined in Table 7 Model Performance without tuning.

Based on the performance of these models, we shortlist models that have R2 > 0.99 for further tuning. Shortlisted models based on these criteria are

- Random Forest
- Light Gradient Boosting
- Extreme Gradient Boosting

Comparing the performance of tuned models against base models.

Table 8 Base Models vs Tuned Models shows that base models without any hyperparameter tuning perform better than tuned models. Base models have high R2 and low RMSE than tuned models.

Evaluating each of these base models in detail.

Table 7 Model performance without tuning.

Model	RMSE	R2
Linear Regression	0.0958	0.486
Decision Tree	0.0127	0.9895
Random Forest	0.0077	0.9961
Gradient Boosting	0.0253	0.9641
Light Gradient Boosting	0.0088	0.9951
Extreme Gradient Boosting	0.0085	0.9957
AdaBoost	0.0655	0.7584

Table 8 Base models vs tuned models.

Base model			Tuned model		
Model	**R2**	**RMSE**	**Model**	**R2**	**RMSE**
Random Forest	0.9961	0.0077	Random Forest	0.9955	0.0088
Light Gradient Boosting	0.9951	0.0088	Light Gradient Boosting	0.9947	0.0093
Extreme Gradient Boosting	0.9957	0.0085	Extreme Gradient Boosting	0.927	0.0361

Table 9 Random forest—Hyperparameters.

Hyperparameter	Value	Hyperparameter	Value
bootstrap	TRUE	min_samples_leaf	1
ccp_alpha	0	min_samples_split	2
criterion	"mse"	min_weight_fraction_leaf	0
max_depth	None	n_estimators	100
max_features	"auto"	n_jobs	−1
max_leaf_nodes	None	oob_score	FALSE
max_samples	None	random_state	5992
min_impurity_decrease	0	verbose	0
min_impurity_split	None	warm_start	FALSE

4.1.1 Random forest

Hyperparameters

Table 9 Random Forest—Hyperparameters states hyperparameters used by Random Forest Regressor.

Residuals and prediction error plot

Above Fig. 12—Random Forest—Residuals and Prediction Error Plot shows residuals and prediction error plots for Random Forest regressor. Residuals for train and test data exhibit a similar trend within the range of ±0.05. Prediction errors are in line with the best fit line with a very slight deviation having an R2 of 0.994.

Learning curve plot

Fig. 13—Random Forest—Learning Curve Plot shows the learning curve for Random Forest regressor. We notice that till 4000 instances learning algorithm keeps improving post which it saturates.

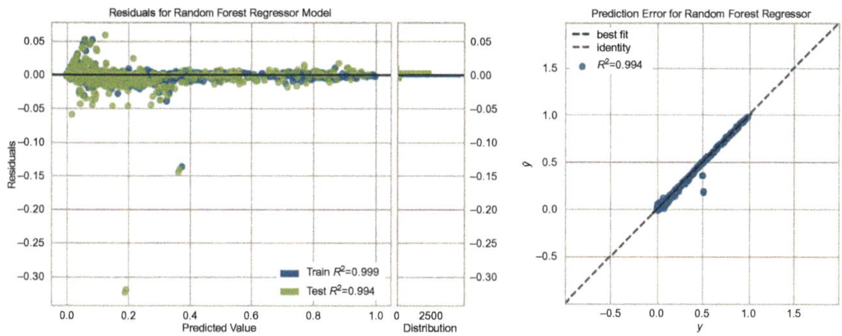

Fig. 12 Random forest—Residuals and prediction error plot.

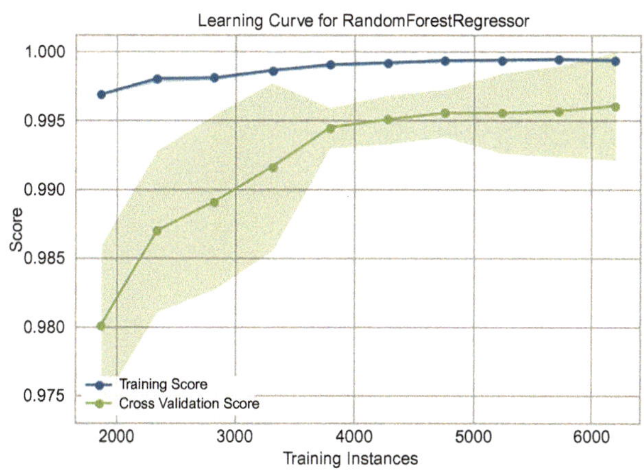

Fig. 13 Random forest—Learning curve plot.

4.1.2 Light gradient boosting

Hyperparameters

Table 10 Light Gradient Boosting—Hyperparameters show the hyperparameters of the model.

Residuals and prediction error plot

Fig. 14—Light Gradient Boosting—Residuals and Prediction Error Plot shows that residuals for train and test datasets exhibit a similar trend within the range of ±0.1. Prediction Errors follow the best fit line with R2 of 0.994.

Learning curve plot

Fig. 15—Light Gradient Boosting—Learning Curve Plot shows a rise in test data score up to 5700 instances post which falls steeply.

Table 10 Light gradient boosting—Hyperparameters.

Hyperparameter	Value	Hyperparameter	Value
boosting_type	"gbdt"	n_jobs	−1
class_weight	None	num_leaves	31
colsample_bytree	1	objective	None
importance_type	"split"	random_state	5992
learning_rate	0.1	reg_alpha	0
max_depth	−1	reg_lambda	0
min_child_samples	20	silent	TRUE
min_child_weight	0.001	subsample	1
min_split_gain	0	subsample_for_bin	200,000
n_estimators	100	subsample_freq	0

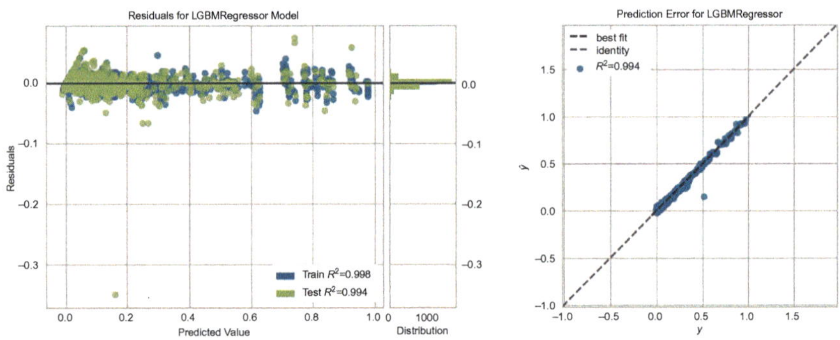

Fig. 14 Light gradient boosting—Residuals and prediction error plot.

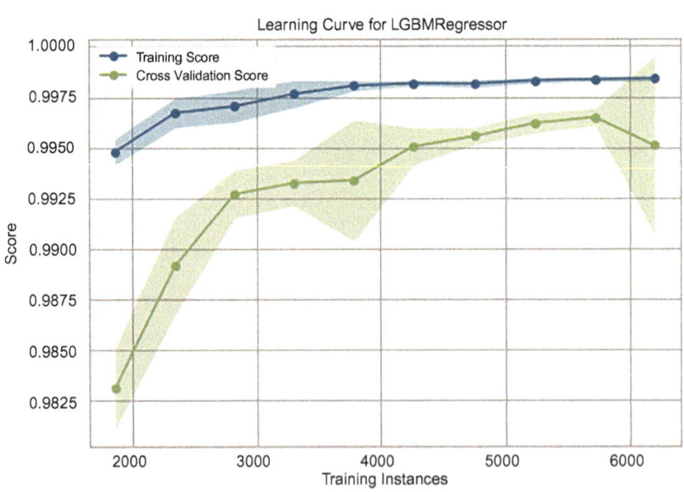

Fig. 15 Light gradient boosting—learning curve plot.

4.1.3 Extreme gradient boosting

Hyperparameters

Table 11 Extreme Gradient Boosting—Hyperparameters show hyperparameters used by the model.

Residuals and prediction error plot

From Fig. 16—Extreme Gradient Boosting—Residuals and Prediction Error Plot, we notice that residuals are limited between the range of ±0.05 for both the train and test dataset. Prediction error follows the best fit line with R2 of 0.993.

Table 11 Extreme gradient boosting—hyperparameters.

Hyperparameter	Value	Hyperparameter	Value
base_score	0.5	monotone_constraints	"()"
booster	"gbtree"	n_estimators	100
colsample_bylevel	1	n_jobs	−1
colsample_bynode	1	num_parallel_tree	1
colsample_bytree	1	objective	"reg:squarederror"
gamma	0	random_state	5992
gpu_id	−1	reg_alpha	0
importance_type	"gain"	reg_lambda	1
interaction_constraints	"	scale_pos_weight	1
learning_rate	0.300000012	subsample	1
max_delta_step	0	tree_method	"exact"
max_depth	6	validate_parameters	1
min_child_weight	1	verbosity	0
missing	nan		

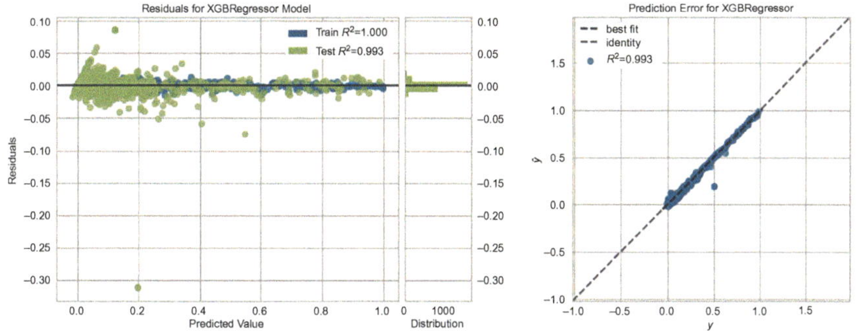

Fig. 16 Extreme gradient boosting—residuals and prediction error plot.

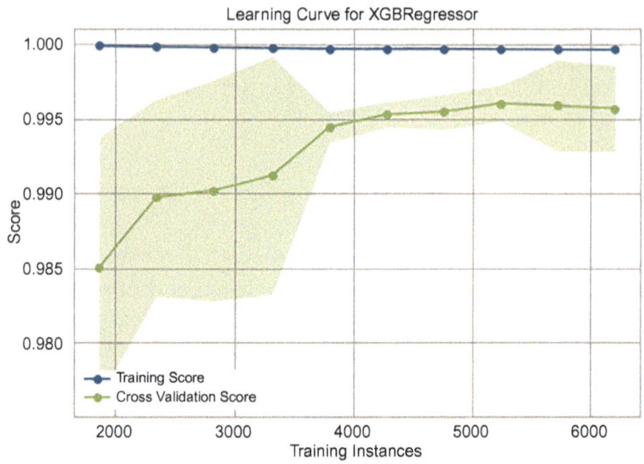

Fig. 17 Extreme gradient boosting—learning curve plot.

Learning curve plot

From Fig. 17—Extreme Gradient Boosting—Learning Curve Plot, it can be deduced that the learning curve increases steeply up to 4000 instances and then saturates.

4.2 Model interpretability

4.2.1 Feature importance

Feature importance is defined by the learning algorithm and is based on the importance of a feature in predicting the target variable.

Random forest

Based on Fig. 18—Feature Importance—Random Forest, we can infer new confirmed COVID-19 cases have maximum impact from confirmed cases in the past 14 days, followed by population density and population. In mobility parameters, transit stations were the highest contributor to new confirmed cases. The basic reproduction rate has a very low impact as compared to other features.

Light gradient boosting

Based on Fig. 19—Feature Importance—Light Gradient Boosting, we can infer new confirmed COVID-19 cases have maximum impact from confirmed cases in the past 14 days, followed by population density and population. In mobility parameters, residential, transit stations, and parks have

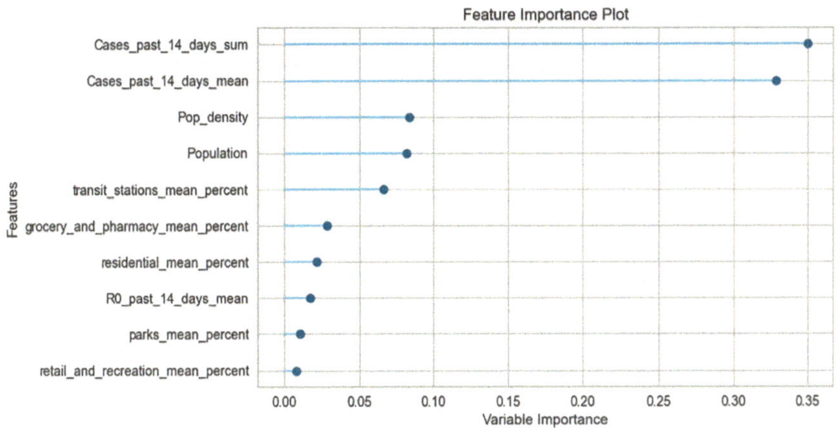

Fig. 18 Feature importance—Random forest.

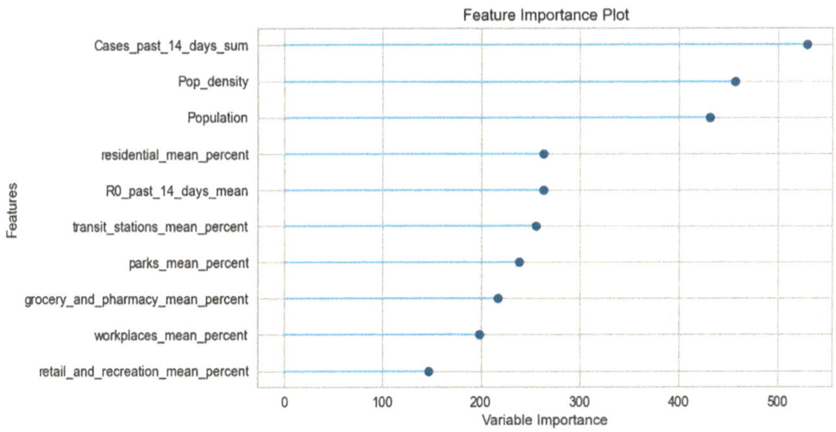

Fig. 19 Feature importance—Light gradient boosting.

good and nearly the same impact, while retail and recreation have the least effect. The basic reproduction rate also has a considerable effect on new confirmed cases.

Extreme gradient boosting

Based on Fig. 20—Feature Importance—Extreme Gradient Boosting, we can infer new confirmed COVID-19 cases have maximum impact from confirmed cases in the past 14 days, followed by population density and population. In mobility parameters, transit stations have a maximum. The basic reproduction rate has a low effect on new confirmed cases.

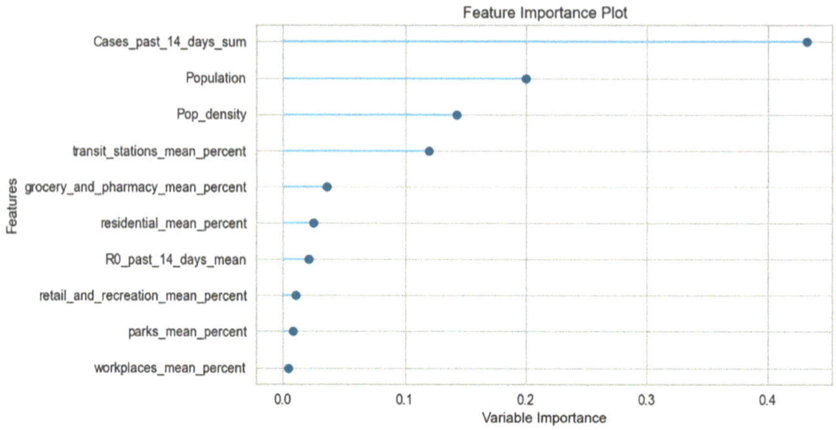

Fig. 20 Feature importance—Extreme gradient boosting.

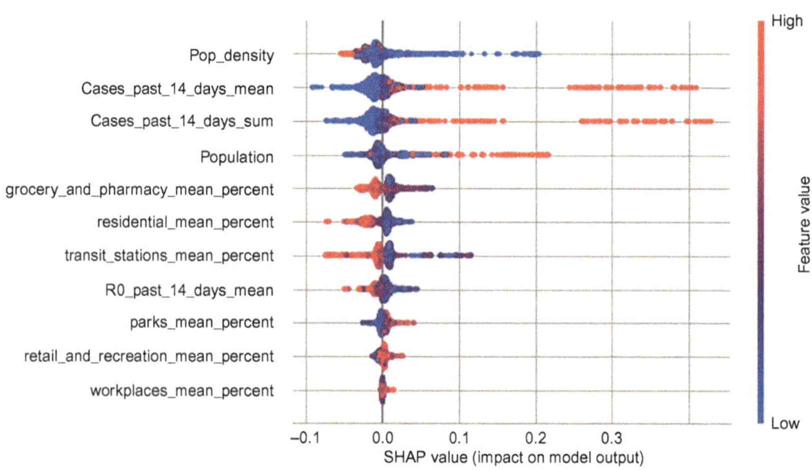

Fig. 21 SHAP interpretability—Random forest.

4.2.2 SHAP interpretability

Random forest

Based on Fig. 21—SHAP Interpretability—Random Forest, we can infer new confirmed COVID-19 cases have maximum impact from confirmed cases in the past 14 days, population density, and population. In mobility parameters, grocery and pharmacy had the most effect on new confirmed cases. The basic reproduction rate has a low impact as compared to other features.

Light gradient boosting

Based on Fig. 22—SHAP Interpretability—Light Gradient Boosting, we can infer new confirmed COVID-19 cases have maximum impact from confirmed cases in the past 14 days followed by population density and population. In mobility parameters, residential and transit stations have a good impact. The basic reproduction rate also has a considerable effect.

Extreme gradient boosting

Based on Fig. 23—SHAP Interpretability—Extreme Gradient Boosting, we can infer new confirmed COVID-19 cases have maximum impact from confirmed cases in the past 14 days followed by population density and population. In mobility parameters, residential and transit stations have a good impact. The basic reproduction rate also has a considerable effect.

4.2.3 LIME interpretability

Random forest

Based on Fig. 24—LIME Interpretability—Random Forest, we can infer a positive impact on new confirmed cases is from the population, followed by mobility parameters—retail and recreation, and transit stations. Cases from the past 14 days have a positive impact along with the basic reproduction rate. The remaining features have a negative effect on new confirmed cases.

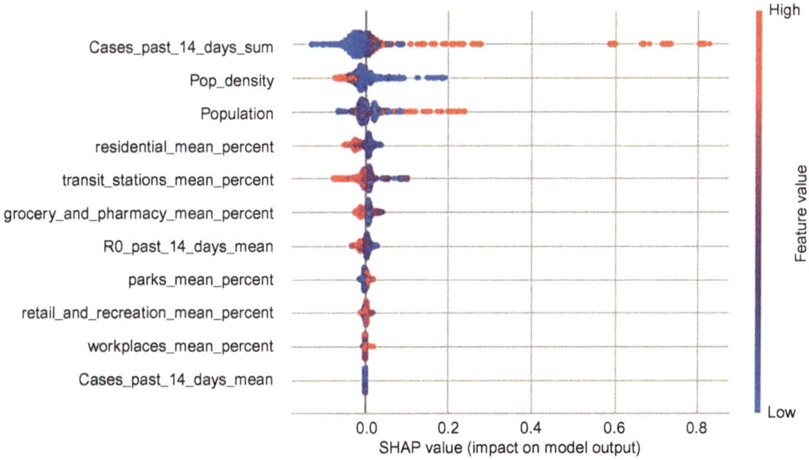

Fig. 22 SHAP interpretability—Light gradient boosting.

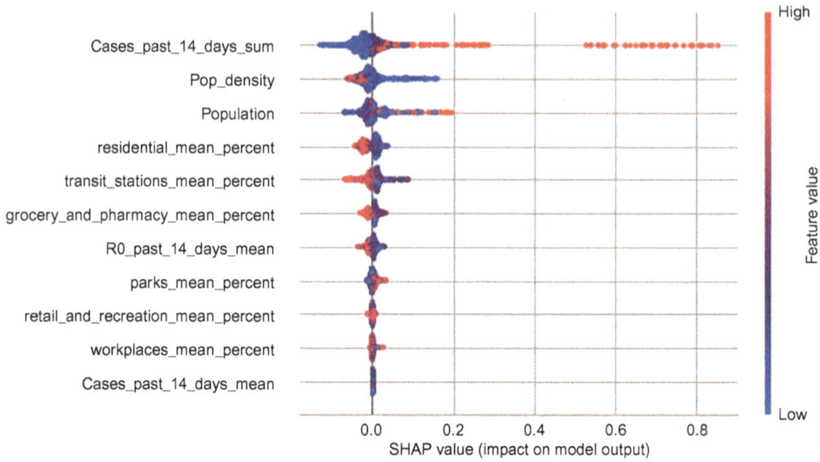

Fig. 23 SHAP interpretability—Extreme gradient boosting.

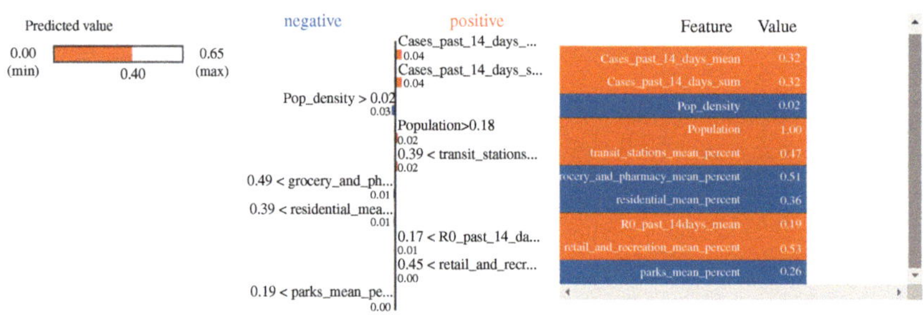

Fig. 24 LIME interpretability—Random forest.

Light gradient boosting

Based on Fig. 25—LIME Interpretability—Light Gradient Boosting, we can infer a positive effect on new confirmed cases is from population followed by mobility parameters—workplaces, retail and recreation, and transit stations. Cases from the past 14 days have a positive impact along with the basic reproduction rate. The remaining features have a negative effect on new confirmed cases.

Extreme gradient boosting

Based on Fig. 26—LIME Interpretability—Extreme Gradient Boosting, we can infer a positive effect on new confirmed cases is from population followed by mobility parameters—retail and recreation, and transit stations.

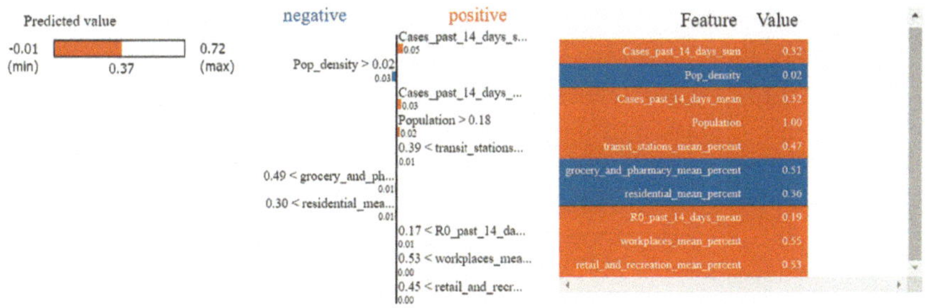

Fig. 25 LIME interpretability—Light gradient boosting.

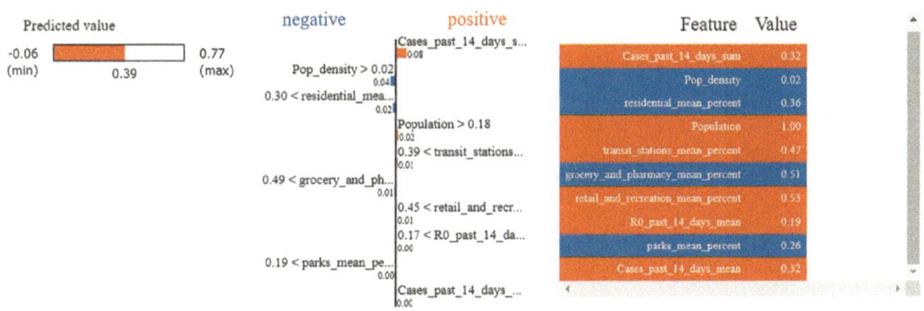

Fig. 26 LIME interpretability—Extreme gradient boosting.

Cases from the past 14 days have a positive impact along with the basic reproduction rate. The remaining features have a negative effect on new confirmed cases.

4.3 Model evaluation

This study is about finding the effect of different features on new confirmed cases, which is done by creating statistical regression models predicting new confirmed cases and comparing the same with actual numbers. Metrics used for model evaluation are Root Mean Square Error (RMSE) and the Coefficient of Determination (R^2).

The following regression statistical regression models are considered for evaluation in this study.

- Linear Regression
- Decision Tree
- Random Forest
- Gradient Boosting
- Light Gradient Boosting
- Extreme Gradient Boosting
- Adaboost

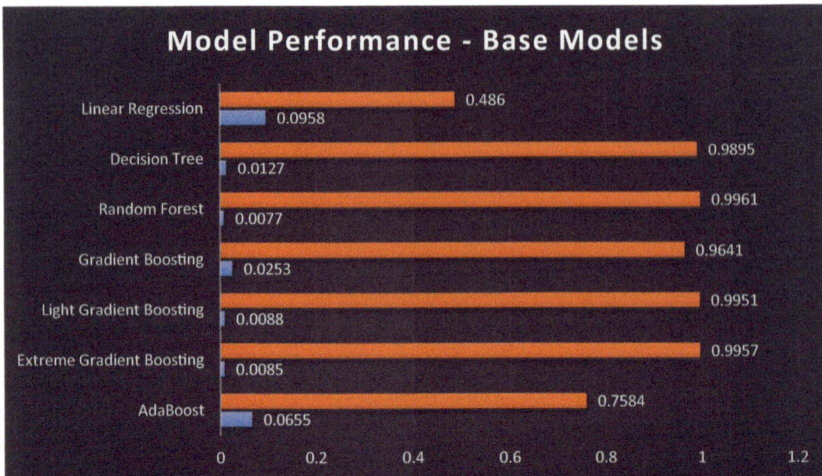

	AdaBoost	Extreme Gradient Boosting	Light Gradient Boosting	Gradient Boosting	Random Forest	Decision Tree	Linear Regression
R2	0.7584	0.9957	0.9951	0.9641	0.9961	0.9895	0.486
RMSE	0.0655	0.0085	0.0088	0.0253	0.0077	0.0127	0.0958

Fig. 27 Model performance—Base models.

Fig. 27—Model Performance—Base Models, we notice high-performing models are Random Forest, Extreme Gradient Boosting, and Light Gradient Boosting, which have similar performance with R^2 in the range of 0.9951–0.9961 and RMSE in the range of 0.0077–0.0088. Followed by Decision Tree with R^2 as 0.9895 and RMSE as 0.0127, and Gradient Boosting with R^2 as 0.9641 and RMSE as 0.0253. The model that performed the worst is Linear Regression with R^2 as 0.486 and RMSE as 0.0958, and AdaBoost with R^2 as 0.7584 and RMSE as 0.0655.

Based on this, further tuning modes are selected with a cut-off on R^2 as 0.99 and shortlisted models are Random Forest, Extreme Gradient Boosting, and Light Gradient Boosting. These models even have the lowest RMSE.

Fig. 28 Model Performance—Base vs Tuned Models compares the performance of models with and without tuning. We notice a dip in performance of all the models post tuning and a major dip is shown by Extreme Gradient Boosting with an approximately 7 points dip in R^2. Random Forest and Light Gradient Boosting after tuning show a dip in performance which is comparable and is not drastic. Based on this observation,

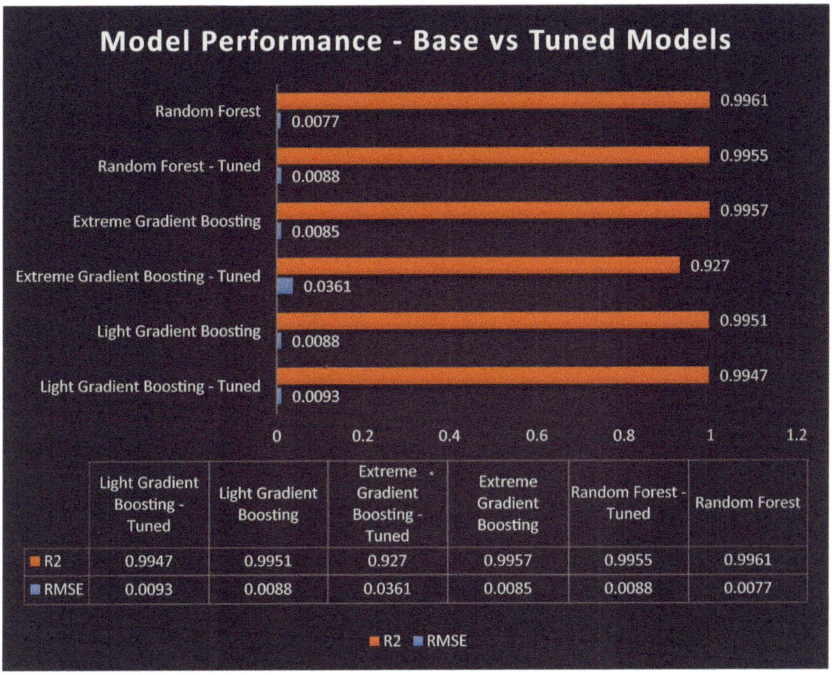

	Light Gradient Boosting - Tuned	Light Gradient Boosting	Extreme Gradient Boosting - Tuned	Extreme Gradient Boosting	Random Forest - Tuned	Random Forest
R2	0.9947	0.9951	0.927	0.9957	0.9955	0.9961
RMSE	0.0093	0.0088	0.0361	0.0085	0.0088	0.0077

Fig. 28 Model performance—Base vs tuned models.

Random Forest, Extreme Gradient Boosting, and Light Gradient Boosting are considered without tuning for model interpretation. In terms of RMSE, all three models have similar performance with a slight deviation. The model that performed best in terms of RMSE and R^2 is Random Forest, with RMSE as 0.0077 and R^2 as 0.9961.

4.4 Model interpretability

As part of model interpretability methods used are feature importance from models, SHAP interpretability, and LIME interpretability. Random Forest, Extreme Gradient Boosting, and Light Gradient Boosting are all interpreted on given interpretability methods. The importance of a feature in explaining target variable new confirmed cases is explained. LIME method predicted values that are in the range of 49–62% of actual values, hence interpretation of LIME could be overridden by other interpretability methods, which make a strong and better case.

Model Interpretability for Random Forest statistical model is shown in Fig. 29—Random Forest—Model Interpretability, Light Gradient Boosting

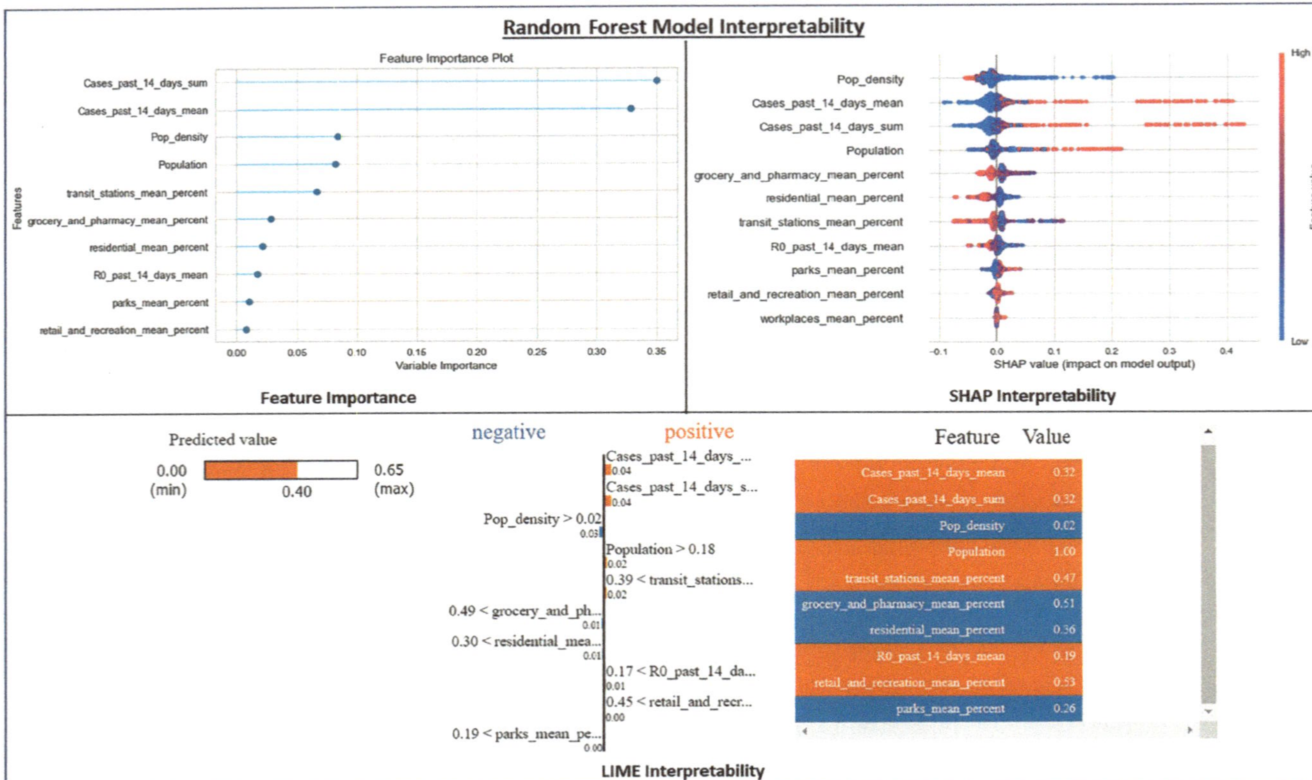

Fig. 29 Random forest—Model interpretability.

in Fig. 30—Light Gradient Boosting—Model Interpretability and Extreme Gradient Boosting in Fig. 31—Extreme Gradient Boosting—Model Interpretability. Model Interpretability against each feature is detailed out below.

4.4.1 Confirmed cases in the last 14 days

From all three statistical models' interpretability, it can be observed that total confirmed cases in the past 14 days have maximum effect on new confirmed cases that is it contributes to an increasing number of new confirmed cases, which is also in agreement with the fact that COVID-19 spreads from person-to-person and can be spread only from existing confirmed cases. Mean confirmed cases in the past 14 days feature has a high correlation with total confirmed cases in the past 14 days; hence it received either similar importance as total confirmed cases or was dropped due to fact of correlation.

4.4.2 Population and population density

It can be observed from all three models' interpretability that the population also contributes to an increase in new confirmed cases with a good magnitude which infers that highly populated locations are prone to have a steep rise in new confirmed cases. For population density, feature importance and SHAP interpretability for all three models depict a high positive impact on new confirmed cases, but LIME interpretability for all three models depicts a slight negative impact that contradicts feature importance and SHAP interpretability. It can also be observed LIME prediction is in the range of 49–62%. Hence we can override LIME observation with feature importance and SHAP interpretability. Based on this, highly populated locations are also prone to a steep increase in new confirmed cases, which is again in alignment with person-to-person transmission of COVID-19.

4.4.3 Basic reproduction rate

The basic reproduction rate positively impacts new confirmed cases from the model interpretability of all three statistical models. So an increase in basic reproduction rate can be deduced as rising in new confirmed cases. This is also in alignment with the definition of basic reproduction rate, according to which it defines the rate at which pandemic is expanding. This feature can be controlled only by curbing the spread of COVID-19.

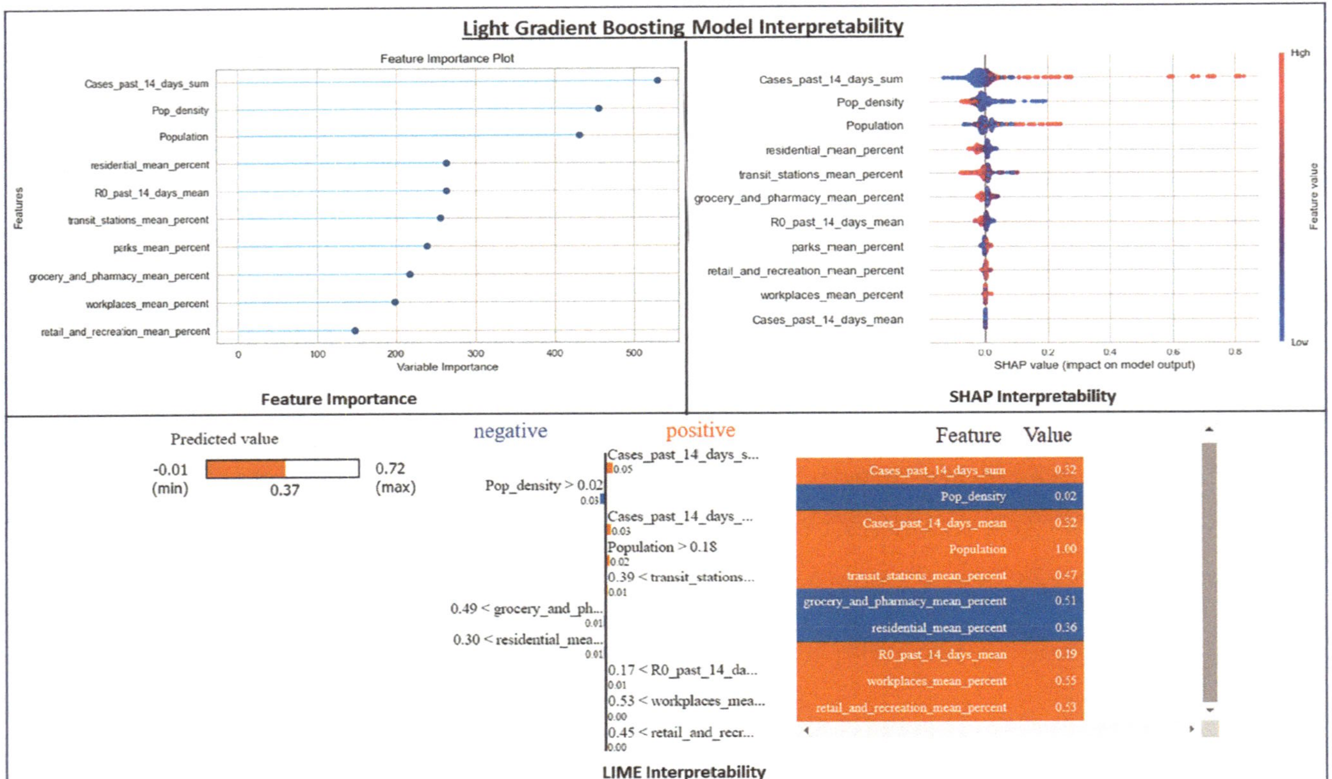

Fig. 30 Light gradient boosting—Model interpretability.

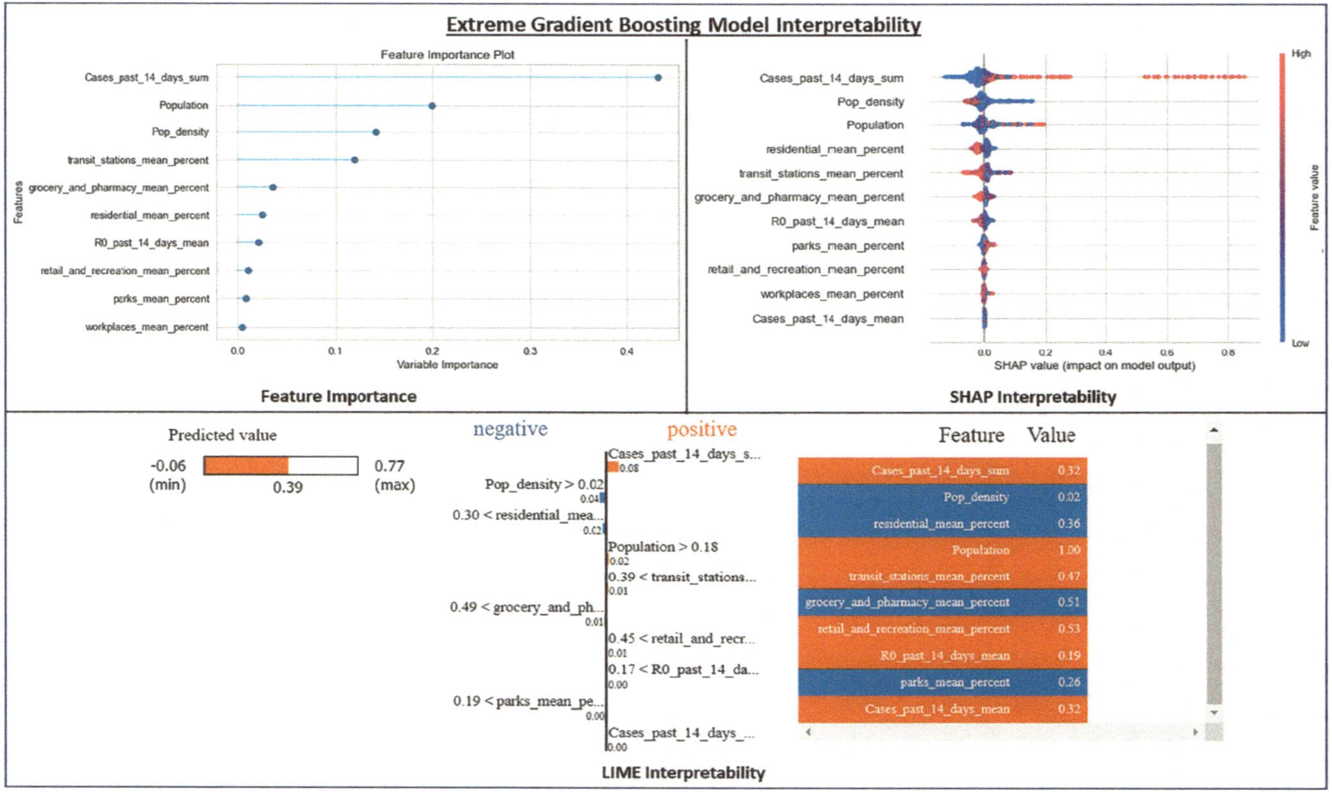

Fig. 31 Extreme gradient boosting—Model interpretability.

4.4.4 Global mobility features

Based on model interpretability from all three models, we have created a formula based on a weighted average to define feature importance.

Assign the value to each feature based on the impact it draws on new confirmed cases ranging from 1 to 6, with 1 being least impact and 6 with maximum impact. Feature value 0 is assigned either feature impact on new confirmed cases is negative, or the interpretability method does not assign any value that drops the feature from calculations. Weights are assigned based on statistical algorithms and interpretability used. Since all three statistical models have similar performance, so weight is going to be 1. For weights about interpretability, feature importance is derived from statistical models; hence it will be given weight as 1, and SHAP is also given weightage as 1. LIME, on the other hand, has a prediction in the range of 49–62% hence is given a weight of 0.5.

Equation is

$$\textit{Net feature effect} = \frac{\sum_{i=1}^{m} M_k \left(\frac{\sum_{j=1}^{n} F_n \times V}{n} \right)}{m}$$

Where,

M = Model which in this case is Random Forest, Light Gradient Boosting, Extreme Gradient Boosting.

F = Model Interpretability method, which is Feature Importance, SHAP, LIME.

m = number of models.

n = number of model interpretability methods.

V = Feature value in the range of features.

From Table 12 Global Mobility Features Importance, it can be inferred that the mobility of people in transit stations is contributing to the new confirmed cases more than all other mobility features. Grocery, pharmacy, and residential places stand second in maximum contribution to new confirmed cases, followed by retail and recreation places and parks. Workplaces stand last, which is since maximum corporate companies have enabled remote working as part of an effort to curb the spread of COVID-19.

Transit stations that top the list for contributing to new confirmed cases need some serious considerations to curb the spread of COVID-19. During onboarding and deboarding times, social distancing has been on the toss, which might be one of the reasons for the spread. Also, the population

Table 12 Global mobility features importance.

Global mobility feature	Random forest			Light gradient boosting			Extreme gradient boosting			Net feature effect
	Feature importance	SHAP interpretability	LIME interpretability	Feature importance	SHAP interpretability	LIME interpretability	Feature importance	SHAP interpretability	LIME interpretability	
Retail and recreation	2	2	6	1	2	4	3	2	5	2
Grocery and pharmacy	5	6	0	3	4	0	5	4	0	3
Transit stations	6	4	5	5	5	6	6	5	6	4
Parks	3	3	0	4	3	0	2	3	0	2
Workplaces	0	1	0	2	0	5	1	1	0	1
Residential	4	5	0	6	6	0	4	6	0	3

dependent on transit stations is relatively high. Grocery and pharmacy is a place where the population approach for daily essentials. During peak time, traffic to such places increases interaction between populations, making them prone to COVID-19. Lately, a lot of family get-togethers are observed, which is again another place of interaction. Parks contributing is significantly less due to the self-isolation meets that can be done in parks due to the massive size of parks.

5 Conclusion and future work

5.1 Discussion and conclusion

This study deals with an ongoing pandemic known as COVID-19 and aims to find non-pharmaceutical interventions that can be implemented in the absence of a vaccine to curb the spread of the virus. An extensive literature review compared to COVID-19 pandemic with previous pandemics and enlightened with different facts about this virus: it spreads from person to person, has an incubation period of up to 14 days, and has a high reproduction rate. The incubation period of COVID-19 is one of the strange aspects to notice, which means a person might not show symptoms of COVID-19 up to 14 days of contraction and in rare cases, this has extended up to 30 days as well. This thing makes controlling the spread of this virus a daunting task. Hence it needs a different approach as compared to previous pandemics to control.

The literature review also discusses vaccine trials, it is found that a typical vaccine has a lifecycle of 18 months with a five-step approach. With the rush to develop a vaccine, steps for the vaccine are accelerated but can't be skipped and with that, the vaccine will take a good amount of time to be available. Post availability of vaccine another set of hurdles of mass production and distribution brings in the next set of challenges. With all of this, curbing COVID-19 in the absence of vaccines becomes a prime objective and steps to keep social and economic balance becomes important. These steps to curb the spread of the virus in the absence of a vaccine are known as Non-pharmaceutical Interventions (NPIs) and the goal of this study is to find NPIs that need to be implemented to curb the spread of COVID-19.

NPIs can be implemented at either personal level, community level, or environmental level and can have the effect of either mitigation (reducing the pace of the spread of pandemic) or suppression (reducing new confirmed cases and sustaining the same). Personal NPIs evolve from self-awareness and responsibility, it includes self-quarantine in case of sick, avoids interaction with people, washing hands, cleaning, or disinfecting frequently touched objects. Community NPIs are general closure or restriction on traffic in public places to reduce population interaction, implement social distancing, and restrict mass gathering. Implementing NPIs curbs the spread of the virus but also brings a good impact on the social and economic conditions of society. Hence NPIs need to be implemented in a balanced way that has maximum effect on curbing the spread of the virus but the least social and economic effect.

This study considers data from the following four different sources.

1. Confirmed COVID-19 cases

   Provides data about new confirmed cases on a daily basis for each US state.

2. COVID-19 community mobility report

   Provides data on the mobility of the population in different places for each US state.

3. US population—States

   Provides data about population and population density for each state in the US.

4. Daily R_0 value—US states

   Provides daily reproduction rate of COVID-19 for each US state.

It further considers only specific feature from each of these data sources to created one integrated dataset for analysis which includes features on daily new confirmed cases, population, population density, daily basic reproduction rate, and global mobility factors for retail and recreation, grocery, and pharmacy, transit stations, parks, workplaces, and residential places. The data is at state granularity on a daily basis and is considered for dates 8th March 2020 to 30th September 2020, both days inclusive. The integrated dataset is taken through multiple pre-processing steps, one the important being smoothening of confirmed cases over ± 2 days to eradicate the weekend effect during the weekend. It was observed low to no cases while Monday used to have an abnormal peak, which was since limited testing was available on the weekend. Outliers were not removed as the study wants to consider all such scenarios. Feature engineering involved considering all features in the past 14 days since the probability of a person found to confirm the case today might have contracted a virus in the last 14 days based on the incubation period.

This is a regression problem aiming to forecast the anticipated number of new COVID-19 cases on a given day in the United States. Based on which relation of a feature on target variable is deduced to define the causal relationship. For this study, stratified K-fold cross-validation with an ensemble of the following models is evaluated.

- Linear Regression
- Decision Tree
- Random Forest
- Gradient Boosting
- Light Gradient Boosting
- Extreme Gradient Boosting
- Adaboost

These models were evaluated based on metrics Root Mean Square Error (RMSE) and Coefficient of Determination (R^2). It was observed Random Forest, Light Gradient Boosting, and Extreme Gradient Boosting performed at par with similar performance while tuned versions of these models had relatively similar performance on the lower side except for Extreme Gradient Boosting, where it dropped significantly. The best performing models on these metrics are Random Forest with RMSE as 0.0077 and R^2 as 0.9961 followed by Extreme Gradient Boosting with RMSE as 0.0085 and R^2 as 0.9957 and Light Gradient Boosting with RMSE as 0.0088 and R^2 as 0.9951.

Further model interpretability explained the effect of each feature on new confirmed cases. Three different methods were used to understand the effect of features on new confirmed cases that were feature importance derived from the model, SHAP interpretability, and LIME interpretability. It was observed that the LIME method predicted values in the range of 49–62%; hence interpretations made by LIME were overridden by other interpretability methods, which made a better and strong case. Based on model interpretability following are the observations.

- *Confirmed cases in the past 14 days*

  It was found that confirmed cases in the past 14 days contribute maximum to new confirmed cases, which are also in alignment with COVID-19 spreads from person to person. This also infers that to curb the spread of the virus, quarantine of confirmed cases needs to follow until recovery strictly. Also contact tracing should be done to tag possible new confirmed cases and a strict quarantine for them as well needs to be ordered until the incubation period is over, along with a negative COVID-19 test.
- *Population and population density*

  It was found that both population and population density contributes to the spread of COVID-19. Hence locations with a high population and population density are more prone to have a steep rise in new confirmed cases. Social distancing should be followed along with the implementation of personal, community, and environmental NPIs.

- *Basic reproduction rate*

  The basic reproduction rate is derived from the pace of the spread of the virus, it showed a positive impact on new confirmed cases. An increase in the basic reproduction rate can be attributed to the high forecast in new confirmed cases.

- *Global mobility features*

  Population mobility in retail and recreation, grocery and pharmacy, transit stations, parks, workplaces, and residential places is considered as part of this study. It was found that mobility in transit stations contributes maximum to new confirmed cases followed by grocery and pharmacy, and residential. Retail and recreation, parks, and workplaces have a very low contribution to new confirmed cases. Hence community and environmental NPIs need to be implemented in transit stations. Frequent cleaning and disinfection, strict social distancing, and restriction of the cluster of people. For groceries and pharmacy delivery and pickup options through the drive-through need to be encouraged. The number of people in-store should be restricted to a basic minimum to follow social distancing. With the oncoming of the holiday, season family gets together are on the rise, for the same personal NPIs should be encouraged to maintain the safety of everyone. Workplaces are already helping curb the spread by pushing remote working, retail, and recreation to use drive-through or open with limited capacity. Parks should also restrict and create small bubbles to avoid mass interactions. Table 13 Conclusion—NPIs identified, summarises NPIs that need to be implemented as part of this study to supress and mitigate the spread of COVID-19.

5.2 Contribution to knowledge

- *Strategy to find NPIs to mitigate and suppress COVID-19*

  This study defines the process of converting data insights into actionable items of implementing which NPI will yield maximum benefit in terms of curbing the spread of COVID-19. The process is based on data available in the public domain and then transform the data and engineer new features based on domain knowledge and factors that define

Table 13 Conclusion—NPIs identified.

Situation/place	NPI
Existing confirmed cases	Stringent quarantine until recovery with a negative test result
Person with possible exposure to the virus	Quarantine up to 14 days with negative COVID-19 test
Places with high or dense population	Personal and community NPIs—includes self-awareness, social distancing, restriction on mass gathering
Transit stations	Social distancing, restrict peak hour rush, restrict entry to only for important use
Grocery and pharmacy	Social distancing, restrict on maximum occupancy, promote drive through pickups or delivery
Residential	Personal NPIs, promote self-awareness
Retail and recreation	Personal and Community NPIs—includes self-awareness, social distancing, restriction on mass gathering
Parks	Personal and Community NPIs—includes self-awareness, social distancing, restriction on mass gathering
Workplaces	Sustain remote working till vaccine is available

COVID-19. Once high influential factors are identified, corresponding NPIs can be implemented to curb the spread of the virus. This strategy helps mitigate the spread of the virus to reduce peak healthcare demand and suppress the virus until the vaccine is available.

- *Strategy to find places that are more susceptible to influence the spread of COVID-19*

  Based on global mobility factors, it defines the strategy of finding the effect of population mobility in a specific place to new confirmed cases. It was also observed that places with a high population and population density are more prone to a steep rise in new confirmed cases. Based on these two observations, strategy or NPI to restrict population mobility and interaction in places that are more prone to contribute to new confirmed cases can be identified.
- *Weighted average to find net feature effectiveness-based interpretability from multiple methods*

  This study used three different model interpretability methods to find feature importance to target variables and then defined a method of inferring net feature importance based on the weighted average.

Weights were defined to statistical algorithms used based on performance and model interpretability methods based on confidence and relevance of the same. This strategy will help integrate interpretability received from multiple sources and define net feature importance.

5.3 Future recommendations

- *Use testing effectiveness as one of the features to define new confirmed cases*
  Testing is one of the key factors that define existing confirmed cases, and adequate testing will increase relevance while gaps in testing are going to create voids. Considering testing data is going to increase visibility on existing confirmed cases and deviation of the same from actual numbers. This will further narrow down NPIs that need to be implemented for effective control of COVID-19.
- *Define NPIs at the state level and compare them against country level*
  The current study uses data at the country level, take data one level deep is going defining a more personalized and focused strategy for a state taking into account state-focused limitations. The densely populated state has a different set of challenges as compared to the sparingly populated state. For example, the pace of the spread of COVID-19 in the densely populated state is high, whereas testing is a limitation in the sparingly populated state.

Credit authorship contribution statement

Suvajit Mukhopadhyay: Supervision and review.
Victor Chang: Review and guideline.

References

Abdel-Basset, M., Chang, V., & Nabeeh, N. A. (2021). An intelligent framework using disruptive technologies for COVID-19 analysis. *Technological Forecasting and Social Change, 163*, 120431.

Abdel-Basset, M., Nabeeh, N. A., El-Ghareeb, H. A., & Aboelfetouh, A. (2020). Utilising neutrosophic theory to solve transition difficulties of IoT-based enterprises. *Enterprise Information Systems, 14*, 1304–1324.

Alloghani, M., Al-jumeily, D., Hussain, A., Aljaaf, A. J., Mustafina, J., & Petrov, E. (2018). Healthcare services innovations based on the state of the art technology trend industry 4.0.2018. In *11th int. conf. dev. esystems eng* (pp. 64–70).

Cao, Z., Zhang, Q., Lu, X., Pfeiffer, D., Jia, Z., Song, H., et al. (2020). *Estimating the effective reproduction number of the 2019-nCoV in China, medRxiv*. Cold Spring Harbor Laboratory Press.

Chen, N., Zhou, M., Dong, X., Qu, J., Gong, F., Han, Y., et al. (2020). Epidemiological and clinical characteristics of 99 cases of 2019 novel coronavirus pneumonia in Wuhan, China: A descriptive study. *Lancet*, *395*, 507–513.

Cheng, H.-Y. Y., Jian, S.-W. W., Liu, D.-P. P., Ng, T.-C. C., Huang, W.-T. T., & Lin, H.-H. H. (2020). Contact tracing assessment of COVID-19 transmission dynamics in Taiwan and risk at different exposure periods before and after symptom onset. *JAMA Internal Medicine*, *180*, 1156–1163.

Coronavirus (COVID-19) events as they happen [WWW Document], n.d. https://www.who.int/emergencies/diseases/novel-coronavirus-2019/events-as-they-happen (Accessed 3 July 2020).

Coronavirus (COVID-19) frequently asked questions | CDC [WWW Document], n.d. https://www.cdc.gov/coronavirus/2019-ncov/faq.html (Accessed 13 June 2020).

Coronavirus Pandemic Data Explorer—Our World in Data [WWW Document], n.d. https://ourworldindata.org/coronavirus-data-explorer?yScale=log&zoomToSelection=true&time=2020-01-22..2020-06-25&casesMetric=true&dailyFreq=true&aligned=true&smoothing=7&country=USA~GBR~CAN~BRA~AUS~IND~DEU~FRA~MEX~CHL~ZAF~DZA~COL&pickerMetric=location&pic (Accessed 25 June 2020).

Coronavirus Vaccine Tracker—The New York Times [WWW Document], n.d. https://www.nytimes.com/interactive/2020/science/coronavirus-vaccine-tracker.html (Accessed 14 June 2020).

Delamater, P. L., Street, E. J., Leslie, T. F., Yang, Y. T., & Jacobsen, K. H. (2019). Complexity of the basic reproduction number (R0). *Emerging Infectious Diseases*, *25*, 1–4.

Dietz, K. (1993). The estimation of the basic reproduction number for infectious diseases. *Statistical Methods in Medical Research*, *2*, 23–41.

Ferguson, N. M., Cummings, D. A. T., Fraser, C., Cajka, J. C., Cooley, P. C., & Burke, D. S. (2006). Strategies for mitigating an influenza pandemic. *Nature*, *442*, 448–452.

Ferguson, N. M., Laydon, D., Nedjati-Gilani, G., Imai, N., Ainslie, K., Baguelin, M., et al. (2020). *Impact of non-pharmaceutical interventions (NPIs) to reduce COVID-19 mortality and healthcare demand* (pp. 3–20). Imperial.Ac.Uk.

Gorbalenya, A. E., Baker, S. C., Baric, R. S., de Groot, R. J., Drosten, C., Gulyaeva, A. A., et al. (2020). The species severe acute respiratory syndrome-related coronavirus: Classifying 2019-nCoV and naming it SARS-CoV-2. *Nature Microbiology*, *5*(4), 536–544.

Gupta, R., Shukla, A., & Tanwar, S. (2020). BATS: A blockchain and AI-empowered drone-assisted Telesurgery system towards 6G. *IEEE Transactions on Network Science and Engineering*. 1–1.

Hellewell, J., Abbott, S., Gimma, A., Bosse, N. I., Jarvis, C. I., Russell, T. W., et al. (2020). Feasibility of controlling COVID-19 outbreaks by isolation of cases and contacts. *The Lancet Global Health*, *8*, e488–e496.

Howe, R. D., & Matsuoka, Y. (1999). Robotics for surgery. *Annual Review of Biomedical Engineering*, *1*, 211–240.

Huang, C., Wang, Y., Li, X., Ren, L., Zhao, J., Hu, Y., et al. (2020). Clinical features of patients infected with 2019 novel coronavirus in Wuhan, China. *Lancet*, *395*, 497–506.

Imai, N., Cori, A., Dorigatti, I., Baguelin, M., Donnelly, C. A., Riley, S., et al. (2020). *Report 3: Transmissibility of 2019-nCoV* (pp. 1–6). Imperial-College-COVID19-transmissibility-25-01-2020.

Lauer, S. A., Grantz, K. H., Bi, Q., Jones, F. K., Zheng, Q., Meredith, H. R., et al. (2020). The incubation period of coronavirus disease 2019 (COVID-19) from publicly reported confirmed cases: Estimation and application. *Annals of Internal Medicine*, *172*, 577–582.

Li, Q., Guan, X., Wu, P., Wang, X., Zhou, L., Tong, Y., et al. (2020). Early transmission dynamics in Wuhan, China, of novel coronavirus–infected pneumonia. *The New England Journal of Medicine*, *382*, 1199–1207.

Liu, Y., Gayle, A. A., Wilder-Smith, A., & Rocklöv, J. (2020). The reproductive number of COVID-19 is higher compared to SARS coronavirus. *Journal of Travel Medicine*, *27*, 1–4.

Liu, T., Hu, J., Xiao, J., He, G., Kang, M., Rong, Z., et al. (2020). Time-varying transmission dynamics of novel coronavirus pneumonia in China. *bioRxiv*. 2020.01.25.919787.

Management of patients with confirmed 2019-nCoV | CDC [WWW Document], n.d. https://www.cdc.gov/coronavirus/2019-ncov/hcp/clinical-guidance-management-patients.html (Accessed 27 June 2020).

Majumder, M., & Mandl, K. D. (2020). Early transmissibility assessment of a novel coronavirus in Wuhan, China. *SSRN Electron Journal*. https://dx.doi.org/10.2139/ssrn.3524675.

Manto, M., Dupre, N., Hadjivassiliou, M., Louis, E. D., Mitoma, H., Molinari, M., et al. (2020). Management of Patients with cerebellar Ataxia during the COVID-19 pandemic: Current concerns and future implications. *Cerebellum*, *19*, 562–568.

Nonpharmaceutical Interventions (NPIs) | CDC [WWW Document], n.d. https://www.cdc.gov/nonpharmaceutical-interventions/index.html (Accessed 20 June 2020).

Moustakas, A. http://arxiv.org/abs/2006.00971.

Ong, S. W. X., Tan, Y. K., Chia, P. Y., Lee, T. H., Ng, O. T., Wong, M. S. Y., et al. (2020). Air, surface environmental, and personal protective equipment contamination by severe acute respiratory syndrome coronavirus 2 (SARS-CoV-2) from a symptomatic patient. *JAMA, the Journal of the American Medical Association*, *323*(16), 1610–1612.

Pan, L., Mu, M., Yang, P., Sun, Y., Wang, R., Yan, J., et al. (2020). Clinical characteristics of COVID-19 patients with digestive symptoms in Hubei, China: A descriptive, cross-sectional, multicenter study. *The American Journal of Gastroenterology*, *115*, 766–773.

Read, J. M., Bridgen, J. R. E., Cummings, D. A. T., Ho, A., & Jewell, C. P. (2020). Novel coronavirus 2019-nCoV: Early estimation of epidemiological parameters and epidemic predictions. *medRxiv*. 2020.01.23.20018549.

Riou, J., & Althaus, C. L. (2020). Pattern of early human-to-human transmission of Wuhan 2019-nCoV. *bioRxiv*. 2020.01.23.917351.

Rt: Effective Reproduction Number [WWW Document], n.d. https://rt.live/ (Accessed 16 August 2020).

Sanche, S., Lin, Y. T., Xu, C., Romero-Severson, E., Hengartner, N., & Ke, R. (2020). High contagiousness and rapid spread of severe acute respiratory syndrome coronavirus 2. *Emerging Infectious Diseases*, *26*, 1470–1477.

Shen, M., Peng, Z., Xiao, Y., & Zhang, L. (2020). Modelling the epidemic trend of the 2019 novel coronavirus outbreak in China. *bioRxiv*. 2019, 2020.01.23.916726.

Sun, J., He, W. T., Wang, L., Lai, A., Ji, X., Zhai, X., et al. (2020). COVID-19: Epidemiology, evolution, and cross-disciplinary perspectives. *Trends in Molecular Medicine*, *26*, 483–495.

Tang, B., Wang, X., Li, Q., Bragazzi, N. L., Tang, S., Xiao, Y., et al. (2020). Estimation of the transmission risk of 2019-nCov and its implication for public health interventions. *SSRN Electron Journal*. https://doi.org/10.2139/ssrn.3525558.

van den Driessche, P., & Watmough, J. (2008). Further notes on the basic reproduction number. *Journal of Heredity*, 159–178. Springer Verlag.

WHO coronavirus disease (COVID-19) dashboard | WHO Coronavirus Disease (COVID-19) Dashboard[WWW Document], n.d. https://covid19.who.int/ (Accessed 29 August 2020).

WHO Timeline—COVID-19 [WWW Document], n.d. URL https://www.who.int/news-room/detail/27-04-2020-who-timeline- - -covid-19 (Accessed 27 June 2020).

WHO | Middle East respiratory syndrome coronavirus (MERS-CoV) (Ed.). (2020). *WHO*.

WHO | Summary of probable SARS cases with onset of illness from 1st November 2002 to 31st July 2003 (Ed.). (2015). *WHO*.

Wu, J. T., Leung, K., & Leung, G. M. (2020). Nowcasting and forecasting the potential domestic and international spread of the 2019-nCoV outbreak originating in Wuhan, China: A modelling study. *Lancet*, *395*, 689–697.

Wu, Z., & McGoogan, J. M. (2020). Characteristics of and important lessons from the coronavirus disease 2019 (COVID-19) outbreak in China. *JAMA*, *323*, 1239.

Zhao, S., Lin, Q., Ran, J., Musa, S. S., Yang, G., Wang, W., et al. (2020). *Preliminary estimation of the basic reproduction number of novel coronavirus (2019-nCoV) in China, from 2019 to 2020: A data-driven analysis in the early phase of the outbreak*. *92* (pp. 214–217). Elsevier.

CHAPTER 6

Cloud-based data pipeline orchestration platform for COVID-19 evidence-based analytics

Mauro Lemus Alarcon, Roland Oruche, Ashish Pandey, and Prasad Calyam
Electrical Engineering and Computer Science Department, University of Missouri-Columbia, Columbia, MO, United States

1 Challenges in COVID-19 data handling

Accessing massive collections of prior medical literature and handling the ongoing data deluge create challenges for health-care data consumers (e.g., clinicians and researchers) who need to make timely data-driven decisions related to the COVID-19 pandemic response. Current practice still heavily relies on time-consuming and onerous manual methods to search, compile, and select the articles that are relevant for gaining insights to shape outcomes (Ioannidis, Salholz-Hillel, Boyack, & Baas, 2020). The COVID-19 pandemic demands swift actions from researchers and clinicians, and there is a dire need for robust tools to help them manage the datasets in research tasks, and also to enable them collaborate with other experts based on critical evidence (Kricka et al., 2020). The tools also need to be integrated within unified data-sharing platforms that increase accessibility to specialized literature and support data analytics automation to expedite, e.g., search and analysis processes. Even more importantly, the tools need to be accessible in a flexible and scalable manner by utilizing cloud-based deployments with necessary interfaces to integrate open-source tools and health-care social networks.

1.1 Cloud- and AI-based data pipeline platform

Data pipelines are increasingly being used to combine data from multiple sources, allow access to multiple users, and include multiple data analytics tools to orchestrate data collection and processing. To handle such data

Novel AI and Data Science Advancements for Sustainability in the Era of COVID-19
https://doi.org/10.1016/B978-0-323-90054-6.00003-9

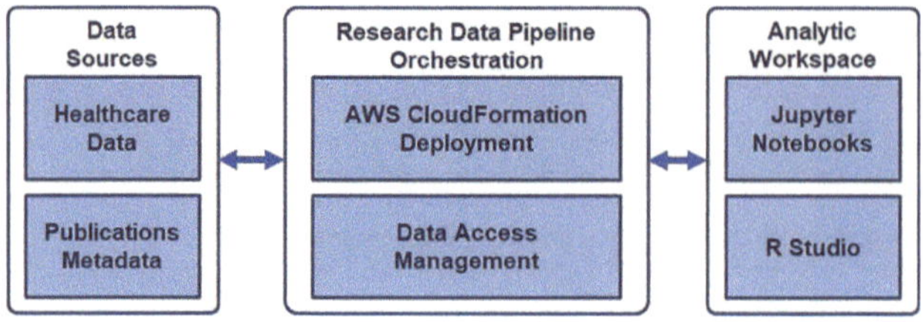

Fig. 1 Cloud-based data pipeline orchestration platform components for COVID-19 evidence-based analytics.

pipelines, exemplar open technologies such as Observational Health Data Sciences and Informatics (OHDSI) (Hripcsak et al., 2015) have been developed, which are yet to be customized and explored for the COVID-19 response purposes. The open architecture of such technologies makes it feasible to integrate open-source data analytics tools and create interactive interfaces to perform data processing. In this chapter, we describe "OnTimeEvidence," a cloud-based data pipeline orchestration platform built on OHDSI for COVID-19 evidence-based text (e.g., publications) and data (e.g., electronic health records) analytics as shown in Fig. 1. More specifically, we describe how OnTimeEvidence leverages the concept of a modern data pipeline platform that uses a technology agnostic architecture on the Amazon Web Services (AWS) cloud platform (Wiggins, 2018) and integrates AI-based analytic tools.

Both structured and unstructured data from multiple sources [i.e., Synthetic Public Use Files (SynPUF) (Borton et al., 2010) health-care data, Kaggle COVID-19 Open Research Dataset (CORD-19) (Ekin Eren, Solovyev, Raff, Nicholas, & Johnson, 2020)] can be stored in a repository that uses redshift data-warehouse services in AWS, and follows the Common Data Model (CDM) standard (Makadia & Ryan, 2014). The repository can be utilized for multiple data processing tasks involving, e.g., natural language processing, machine learning depending on assigned roles of users who have relevant entitlements for managing data access and processing. Users manage the data processing tasks in their customized analytics workspaces that provide the necessary tools to retrieve and analyze the data with user-friendly Jupyter Notebooks and R Studio interfaces. A particular open-source tool that we integrate in OnTimeEvidence is the Domain-specific Topic Model (DSTM)-based publication analytics tool (Zhang, Calyam, Joshi, Nair, & Xu, 2018) that helps users in inferring latent patterns across COVID-19 or other scientific domain literature documents. Using a detailed case study, we show how OnTimeEvidence helps health-care data consumers to submit data requests and retrieve the data in a secure, consistent, and standard manner to analyze COVID-19-related literature. Users are also provided with access to analytic tools which help them to (a) conduct knowledge discovery tasks while reducing the manual burden in compiling and analyzing large datasets and (b) run data analytic processes on disparate systems with minimal automation.

1.2 Chapter organization

In this book chapter, we first present the background of open technologies and issues around cloud-hosted data processing pipelines with AI-based tools. Next, we introduce our OnTimeEvidence platform and detail its architecture components. Following this, we present a case study to show the benefits of deploying OnTimeEvidence to help with COVID-19-related data analytics requiring data access control and use of versatile analytic workspaces. Lastly, we list a set of open issues for how cloud- and AI-based platforms could be further developed to not only handle COVID-19 crisis, but also support needs of health-care data consumers to unlock the promise of "precision medicine" that can help better cure cancer and other diseases (Friedman, Letai, Fisher, & Flaherty, 2015).

2 Background and related works

In this section, we first provide a background on existing exemplar cloud-based data pipeline orchestration solutions (i.e., the OHDSI on AWS platform) that motivate our OnTimeEvidence platform design. Next, we describe best practices in cloud-based health-care data analytics and sharing. Lastly, we summarize latest advances in cloud-based data processing pipeline schemes that can benefit health-care data consumers.

2.1 OHDSI on AWS infrastructure

The OHDSI program is committed to promote the importance of health data analytics through the development and release of open-source data analytics tools (i.e., ATLAS, ACHILLES, ATHENA) (Hripcsak et al., 2015). These tools have common features which allow them to interact with a CDM (Makadia & Ryan, 2014) that can be implemented using multiple database management systems (e.g., Postgresql, Redshift). Through proper extraction, transformation, and loading (ETL) processes, disparate structured and unstructured data sources can be integrated into the CDM repository under a well-defined data structure that will allow the analytic tools to utilize templates to run standardized data analytic processes and generate insightful results.

Our OnTimeEvidence platform builds on the open-source OHDSI on AWS solution, and extends the out-of-the-box automated CloudFormation deployment that includes a Redshift data-warehouse infrastructure instance. This instance hosts the CDM repository physical model, and the data analytic tools that allow OnTimeEvidence users to interact with the CDM. As part of the OnTimeEvidence deployment, we have loaded the SynPUF dataset into the CDM. The deployment of this platform takes a few hours but it removes the manual burden in the design, development, and deployment efforts required by a regular IT infrastructure process with manual steps. We complemented the data repository by adding tables to store the CORD-19 metadata about COVID-19 literature, and loaded the related dataset into those tables. On top of this infrastructure, we developed a centralized role-based data access model to provide entitlements to authorize users to access both datasets and analytic resources. To facilitate the data retrieval and analysis tasks, we developed embedded data request forms within a JupyterLab environment that enables researchers to submit data requests and retrieve the required data within the Jupyter workspace. Once the user-requested data is available, users have access to various analytic tools, and can run correlation rules on multiple datasets. Thereby, they can produce relevant diagnostics to discover insights for COVID-19-related research tasks.

2.2 Cloud-based health-care data management

Multiple solutions have been developed to store and share health-care data in cloud environments, keep those records secure in such environments, provide analytic services related to health big data, and preserve data privacy. In the context of data accessibility, the work in Health-care Data Gateway (Yue, Wang, Jin, Li, & Jiang, 2016) aims to securely store Electronic Health Record (EHR) data in a cloud-based platform and uses a Blockchain-based secure storage layer. Data sharing is supported among multiple users (i.e., physicians, researchers, government institutions, private organizations) based on role assignments. Similarly, in their work, Matos, Pardal, Adao, Silva, and Correia (2018) proposed a system to store EHR in a public cloud, and their focus was on ensuring data confidentiality and integrity by using an access control mechanism based on the lattice model. Using such a model, users can define a hierarchy of data access levels (i.e., private, clinic, research, public). Identify and access management solutions focusing more on cloud-based data sharing while preserving privacy as have been proposed by Hörandner, Krenn, Migliavacca, Thiemer, and Zwattendorfer (2016), Sharma, Chen, and Sheth (2018), and Barik, Dubey, and Mankodiya (2017). In these works, once a user authentication is complete, data access from multiple client devices can be allowed or patient data can be collected from multiple sources. However, none of these prior works provide user-customized analytics workspaces, which in turn leads to users having to manage any retrieved data manually outside of their platforms.

Access management best practices related to centralization need to be carefully designed (Cohen & Nissim, 2018). Among the access management best practices, role-based access control (RBAC) has been the most popular one (Dinakarrao et al., 2019). RBAC restricts platform users' permissions to their roles and only permits users access to privileges that they absolutely need to perform their job functions. For example, health-care students of an organization should not have access to digital financial records of patients. In addition, RBAC also helps facilitate identity security, operational processes, and cybersecurity visibility. As part of our access management best practices, it is important to assign clear, delineated roles to all users. Ideally, this includes privileged users such as faculty members with Institutional Review Board (IRB) approved projects having more entitlements compared to regular users such as students. Moreover, RBAC implementations need to ensure that no user should receive permissions outside his/her role. However, if projects demand the assignment of temporary privileges, those privileges should expire within a set time limit to ensure long-term security of the data access.

2.3 Cloud-based data processing pipelines

Prior works have exemplified the need to provide open-source, cloud-based frameworks for the deployment of data processing pipelines. In the work by García et al. (2020) a distributed cloud-based framework has been developed, viz., DEEP to enable researchers to process and train their machine learning data science models. The DEEP framework integrates serverless architecture to ease the transition from deployment to production. In a similar fashion, VariantSpark (Bayat et al., 2020) is a distributed machine learning framework that performs association analysis to effectively identify variants with complex phenotypes. It features a multilayer parallelization that allows the framework to scale the whole genome population dataset for developing an in-depth analysis using its machine learning pipelines. Simmhan et al. (2013) resolved the critical concern of optimizing supply-demand needs of customers in a Smart Grid Project by developing a robust cloud framework that leverages machine learning and data processing pipelines.

In recent work related to the COVID-19 pandemic, Tuli, Tuli, Tuli, and Gill (2020) resolved the need to handle increasing rate of COVID-19 through a data-driven model deployed on a cloud-based framework that predicts the growth of the pandemic. Abdel-Basset, Chang, and Nabeeh (2020) developed an intelligent framework of emerging AI-based technologies for helping with the COVID-19 pandemic response. Their work suggests that these disruptive technologies can be integrated in IoT and IoMT devices using cloud platforms. Abdel-Basset, Chang, & Mohamed, 2020 sought to resolve the image segmentation problem in COVID-19 chest X-rays by developing a novel machine learning framework that utilizes slime mold and whale optimization algorithms. This problem involves a threshold mechanism that builds a binomial classification to determine whether a patient has the COVID-19 virus. Similarly, Abdel-Basset, Chang, Hawash, Chakrabortty, and Ryan (2021) addressed the issue of providing accurate classification of COVID-19 in CT scans by developing a deep learning architecture that leverages a semisupervised few-shot segmentation algorithm for image segmentation. The work by Otoom, Otoum, Alzubaidi, Etoom, and Banihani (2020) presents an Internet of things (IoT)-based framework that performs real-time monitoring and tracking of COVID-19 data. The framework entails the aggregation of data from multiple resources in a cloud infrastructure where stakeholders (e.g., health physicians) can monitor patients through data processing algorithms. A study by Ashraf et al. (2020) developed a smart surveillance system for effective remote monitoring of human health conditions and close interactions. Similarly, Hossain, Muhammad, and Guizani (2020) developed a mass surveillance system through a hierarchical edge computing service using 5G wireless connectivity and deep learning algorithms.

OnTimeEvidence builds upon the above works and uses data-driven models deployed on OHDSI to allow researchers to effectively perform knowledge discovery pertinent to the COVID-19-related datasets. We develop a data-driven modeling scheme through our existing Domain-specific Topic Model (DSTM) (Zhang et al., 2018), which is an extension of the Latent Dirichlet allocation (Blei, Ng, & Jordan, 2003) to discover the relationships between words and tools/resources (e.g., drugs and genes) related to the COVID-19 pandemic. In addition, we also use Gibbs sampling algorithm (Griffiths & Steyvers, 2004) to infer latent patterns within the COVID-19 domain in an unsupervised manner.

3 OnTimeEvidence architecture and component implementation

In this section, we introduce our OnTimeEvidence platform and its components as illustrated in Fig. 2. The core component of the data pipeline orchestration is built on top of the open-source OHDSI on AWS. The AWS CloudFormation is used to deploy OnTimeEvidence along with the data access and process management module, entitlement database, and access admin console. We leverage the JupyterLab included with OHDSI on AWS to facilitate users' data access and interaction, and create extensions such as the user data request forms and the data processing models in order to provide the analytic workspace for health-care data and COVID-19 publications analysis as well as result sharing. We have uploaded the SynPUF Medicare and the CORD-19 datasets to a relational repository on the OHDSI Red-shift data warehouse service, and the related health-care data and COVID-19-related literature information are available for process testing and validation of user utility.

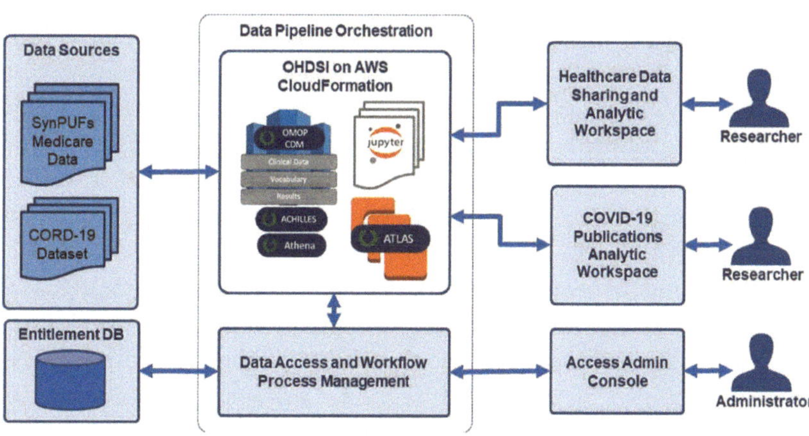

Fig. 2 Components of OnTimeEvidence data pipeline orchestration built on top of the OHDSI on AWS infrastructure.

3.1 OHDSI components of OnTimeEvidence

OnTimeEvidence uses various OHDSI components featuring in-built data repositories and software capabilities. The deployment of OHDSI components provides an enterprise class, multiuser, and scalable health-care data sharing and analytics functionality. As shown in Fig. 2, OHDSI components include the OMOP-CDM deployed on a Redshift data warehouse. The CDM schema allows the integration of disparate data sources into a common format (model) and common representation (terminology, vocabulary, coding) allowing the definition and execution of standard analytic processes. Once data is available in the CDM, evidence knowledge can be generated using the analytic tools included with the OHDSI on AWS platform (i.e., Athena, ATLAS), and the analytic models and tools available in the analytic workspace available via Jupyter Notebooks or R-Studio. The OHDSI components include out-of-the-box open source analytic tools such as (i) ATLAS, a web-based application for researchers to conduct analyses on data loaded to the OMOP-CDM through the creation of cohorts based on drug exposure or diagnosis of a particular condition. The cohort results are visualized in the tool's user interface, or stored in a relational repository to be used by other analytic tools; (ii) ACHILLES, an application used to analyze the database hosting the CDM and evaluate data quality; and (iii) ATHENA, a tool that is used to generate and load standardized data vocabularies into the CDM repository.

OnTimeEvidence extends OHDSI through the integration of the following new components we have developed: (i) a role-based user access and workflow management component to keep control on the authentication and authorization of the data, and ensure data privacy and security compliance; this functionality allows users to submit data requests, which are fulfilled by OnTimeEvidence based on the role-based user access privileges and (ii) the functionality that allows users to perform publication analytics (considering knowledge pattern mining of hundreds or thousands of articles) and big data analytics (considering millions of patient-related records) relevant to COVID-19. The original OHDSI features are complemented by the above two components and the result is a robust platform for researchers to analyze data with open-source tools, and find correlations as well as gain insights all within the same platform. Consequently, OnTimeEvidence reduces the burden of researchers to handle large-scale data compilation and analysis in cloud infrastructures, and enables them to focus on their scientific research goals instead.

3.2 Access and authorization management

Herein, we provide the details regarding the data access and workflow management components in OnTimeEvidence. For user management, we created a centralized role-based access control mechanism to manage users' credentials and privileges through an administrator interface. The user account creation workflow process steps are illustrated in Fig. 3. Users are first asked via the OnTimeEvidence web interface to provide essential data to validate their privileges and necessary data requirements for their analytics tasks. Users are accordingly granted role privileges by the administrator using roles such as student, faculty, or independent researcher depending on the data access requirements and user status in an institution. Once the administrator verifies user credentials, a user account corresponding to a group is created on the OHDSI database server through an administrator web interface that runs customized shell scripts in the backend to automate the user account creation and group mapping. After user account creation, administrators can also dynamically regulate the level of access provided to users thus protecting sensitive and proprietary data. A large number of roles consisting of a combination of access to different attributes can be stored in the entitlement database to allow administrators on creating user accounts with flexibility in access control rules governing the platform.

The control actions performed by the administrator utilize the HTTPS protocol to connect and communicate with the application and database server in a secure manner. Administrator password and user access request form are secured on a separate database system (disjoint with the original

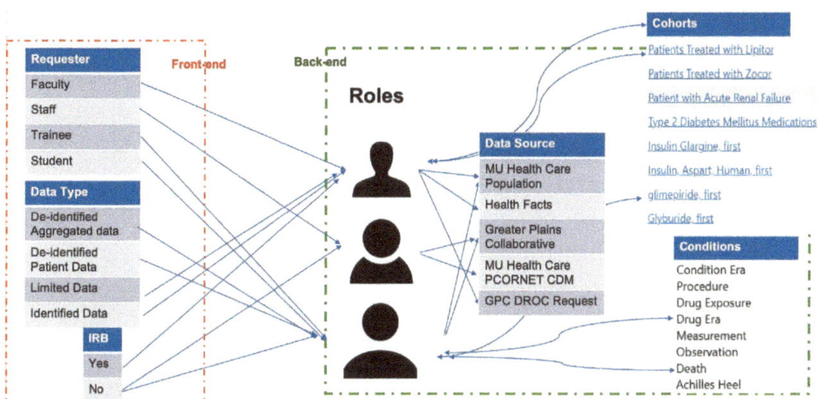

Fig. 3 Role-based access control enabled by a web interface and entitlement database using automated shell scripts.

OHDSI components, e.g., CDM) identified as the entitlement database server in the architecture schematics shown in Fig. 2. The role-based access control and separation of concern for user access request form and administrator authentication on a different entitlement (database) server reduces the attack surface on the overall platform. To increase security, we enforce strong passwords (at least eight characters with uppercase and lowercase letters, numbers, and special characters) for authentication of users and the administrator. We have added further security measures by using the *bcryptjs* library for hashing the password and other user data at rest. Field level encryption is used for protecting sensitive user data such as details about the project or their privileges. This ensures that the actual explicit password and data are not accessible to hackers even if they have access to the entitlement database. To save the users from malicious security attacks, strict protocols such as form and data validations are put in place at each layer of the control flow. Consequently, a malicious user cannot execute attacks such as SQL injection attacks on the application or database server. Data validation at the application layer is also performed to prevent users from storing undesired data on the web server.

3.3 COVID-19 literature selection and analysis

The OnTimeEvidence platform can be used with customized analytics workflows in which health-care data consumers (e.g., clinicians, health professionals) access COVID-19 literature in their scientific research tasks. In clinical fields, researchers commonly follow a systematic literature review procedure known as the evidence-based practice (Sackett, 1997). Health-care data consumers commonly adopt this method for synthesizing and reviewing articles based on the inherent evidence levels that are pertinent to their research. Specifically, a hierarchical evidence-based framework, viz., Levels of Evidence Pyramid (Murad, Asi, Alsawas, & Alahdab, 2016), is used. The Levels of Evidence Pyramid illustrate the reduced quantity of publications with respect to the increase in high-quality information (e.g., background information to systematic reviews). However, it remains a challenge for clinical researchers to sort and filter information based on high-quality evidence in a timely manner.

To simplify literature data selection and analysis for COVID-19 researchers, we implement new components in the OnTimeEvidence platform to cater to the user's needs by reducing their manual steps in scientific workflows as illustrated in Fig. 4. The implemented tasks performed include:

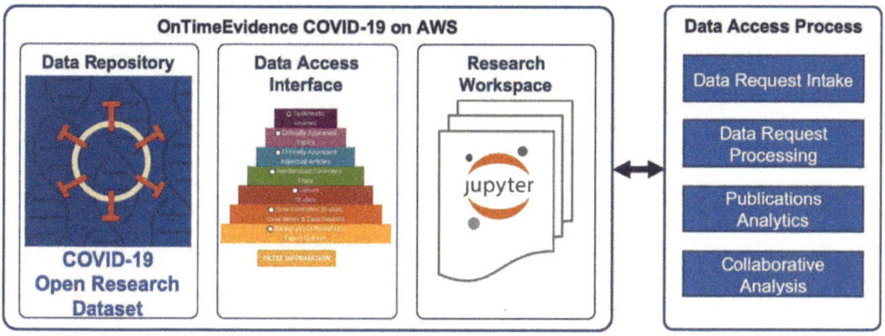

Fig. 4 COVID-19 literature selection and analysis workflow process in OnTimeEvidence for knowledge discovery.

(i) a literature selection form that allows a user to query search terms related to the levels of evidence, (ii) the functionality to process the literature selection, and, in response, generation of a new Jupyter Notebook with an embedded SQL query to execute the information requested by the user, and (iii) use the JupyterLab workspace to allow the user to conduct a publication and/or collaborative analysis and store the results for sharing them via the platform.

The above OnTimeEvidence component for literature selection and analysis described uses a relational data structure and process to upload and store the metadata related to the COVID-19 literature. For the purposes of this work, we have collected over 10,000 publication records from the Kaggle COVID-19 Open Research Dataset (CORD-19) (Ekin Eren et al., 2020). This dataset is stored on an Amazon Redshift database server that hosts CDM data models as shown in Fig. 2. With the publication archives stored in the Red-Shift cluster, the user is able to search for articles by using the request form through a data access request form on a JupyterLab interface. The form allows the user to submit a COVID-19 literature selection based on the selected Level of Evidence. Following this approach, the data access request form processes the health-care data consumers' requests and generates the SQL query to retrieve the related literature selected. In this process, we have developed an intuitive client-interface form that is rendered on the JupyterLab workspace within the default view for data consumers, i.e., the request form is displayed whenever users have been given accesses by the administrator in the workspace, and they can use it to submit a new publications data access request.

Fig. 5 illustrates the process of the CDM to execute the SQL query and generate a new Jupyter Notebook on the user's workspace once the data

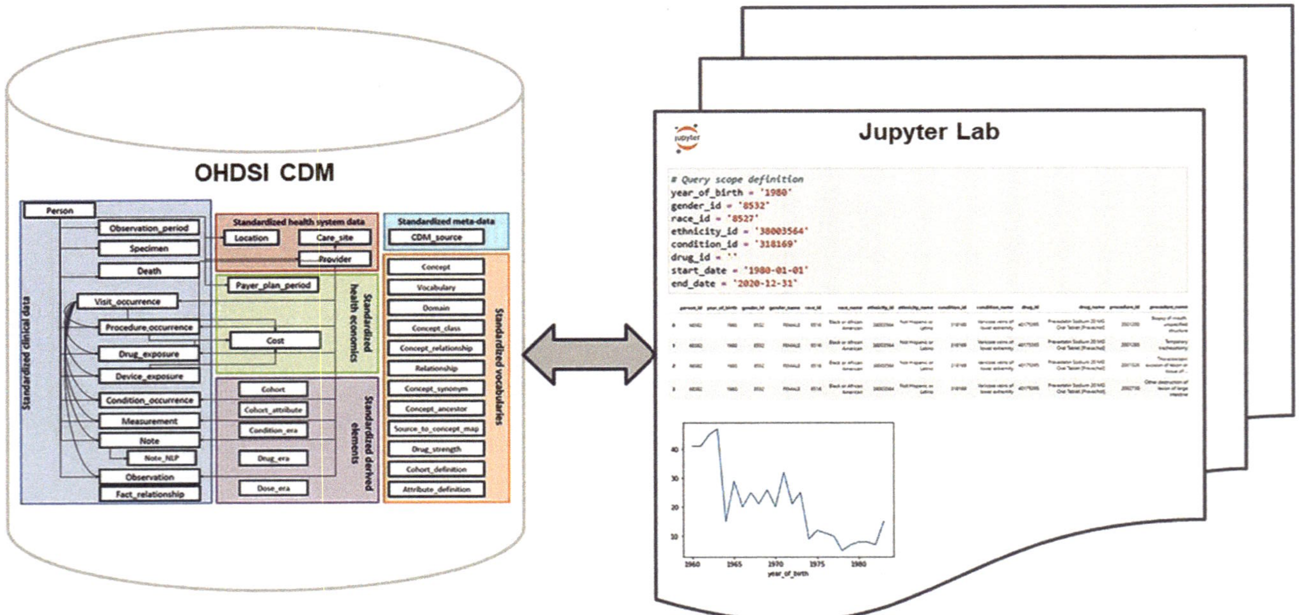

Fig. 5 Integration between a Jupyter Notebook and the CDM via a SQL query to retrieve data and perform data analytics for a given COVID-19 research task.

request form is submitted and processed from the user's query. The user will use this notebook to execute the query against the CDM and retrieve the required data. The user does not have to know the structure or content of the CDM repository to retrieve data as the query provides the required definition to fulfill the data request. However, if the user has some knowledge of the CDM structure and data content, it will be possible for the user to modify the query and retrieve a new dataset. The scope of the data being accessed by the user will be limited by the user's role. Therefore, even if the user attempts to access data not allowed for the related role, the corresponding query will not work due to the access security mechanism. This feature allows the user to explore only the data that the user has access to when using the initial SQL query statement or modify the queries in the user data request submission. New SynPUF data following the CDM model can also be added and processed in the OnTimeEvidence platform for varying analytics challenges by the researchers and administrators. Once the SQL query provides a required list of selected articles, the user will be able to conduct analytic tasks using the Python libraries available in the JupyterLab environment, generate the required results, and store or share those results within the same environment.

3.4 Data processing using domain-specific topic model

With the implementation of new components on top of OHDSI, OnTimeEvidence enables the collection of external resources and deployment of machine learning models via open-source tools. In this section, we illustrate the utility of open-source tools in OnTimeEvidence to filter high-quality information and reduce the time-expensive workflow steps involved in performing knowledge discovery over COVID-19 publication archives using data processing pipelines.

Particularly, we detail our Domain-specific Topic Model (DSTM) (Zhang et al., 2018)-based tool that can be used to deploy statistical and deep generative models for guiding researchers to rapidly discover high-quality information (in terms of Evidence Levels) from aggregated medical resources (e.g., publication databases, information on drugs and genes). DSTM extends the Latent Dirichlet Allocation (LDA) (Blei et al., 2003) model to discover domain-specific topics and latent knowledge patterns among the topics and users' interests. The LDA model is a powerful tool that is capable of representing a document through a Dirichlet distribution of random topics and, in parallel, represents each topic as a distribution of key terms. Our DSTM also utilizes inference algorithms such as variational inference and Gibbs sampling (Griffiths & Steyvers, 2004) to infer those latent parameters. DSTM learns two distinct multinomial distributions (Dai, Ding, Wahba, et al., 2013) to generate random topics: (i) topic distributions that are distributions over terms and (ii) document distributions that also are distributions over topics. Given that LDA generates random topics in the scientific literature, we have configured our DSTM tool to generate domain-specific topics through the aggregation of scientific terms related to the COVID-19 research areas.

In utilizing the LDA model for identifying the distribution of topics and their respective terms, we leveraged our DSTM (Zhang et al., 2018) to discover the latent patterns of specific scientific drug and gene terms in salient medical information from the COVID-19 Vaccine Tracker (Milken Institute, 2020) and Virtual Incident Procurement (ViPR) (Pickett et al., 2012). Clinical researchers commonly refer to important criteria including drugs and gene tools to further study the issues relating to infectious disease, immunology, and epidemic/pandemic control. Hence, we simplify the computational complexity of our generative model by generating each word based on a drug or a gene.

As shown in Fig. 6, our DSTM works automatically learns the latent patterns underlying the datasets. Each document represents the collected document from the CORD-19 Kaggle dataset. We decompose a scientific paper into a list of information about the research topic, and related drugs

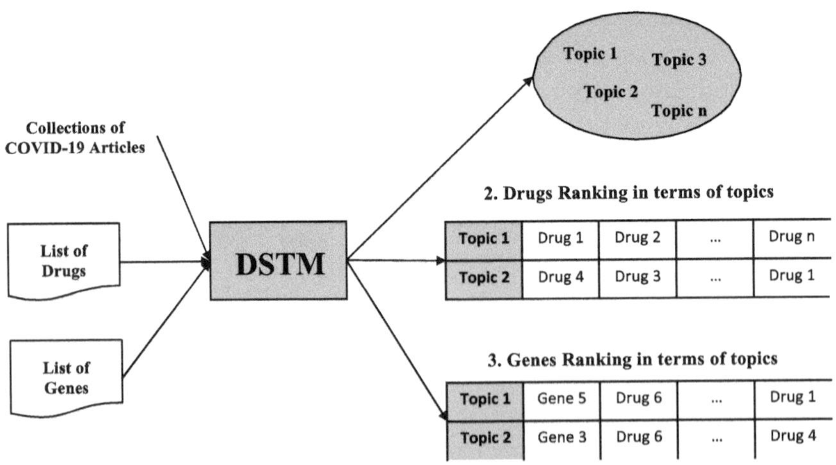

Fig. 6 Domain-Specific Topic Model (DSTM) deployed on OnTimeEvidence works as an analysis engine to discover the relationships among research topics, drugs, and genes.

and genes mentioned in the publication. The goal of DSTM is to learn the relationships among the research topics, drugs, and genes using an unsupervised machine learning approach. To train the DSTM, we only need to input a collection of COVID-19 articles, a list of drugs, and a list of genes. After completion of the training phrase, the DSTM can help users to analyze the most popular research topics in the articles as well as help rank the most commonly investigated drugs or genes based on each topic. Our DSTM can also effectively help scientists query COVID-19 relevant drugs and genes based on their research topics, or search relevant COVID-19 research topics based on specific drugs and genes.

4 OnTimeEvidence COVID-19 case study

Herein, we demonstrate the utility of the OnTimeEvidence platform for COVID-19-related data analytics in the form of a case study. As part of the data access management in OnTimeEvidence, we created multiple user profiles, assigned profiles to users, and ran tests related to data access based on the defined profiles. As shown in Fig. 7, we validated basic security checks, e.g., the web interface for the access request form to create user credentials with necessary validations. Thus we ensured that we prevent SQL injection attacks and fake malicious user(s) creation. We remark that the web

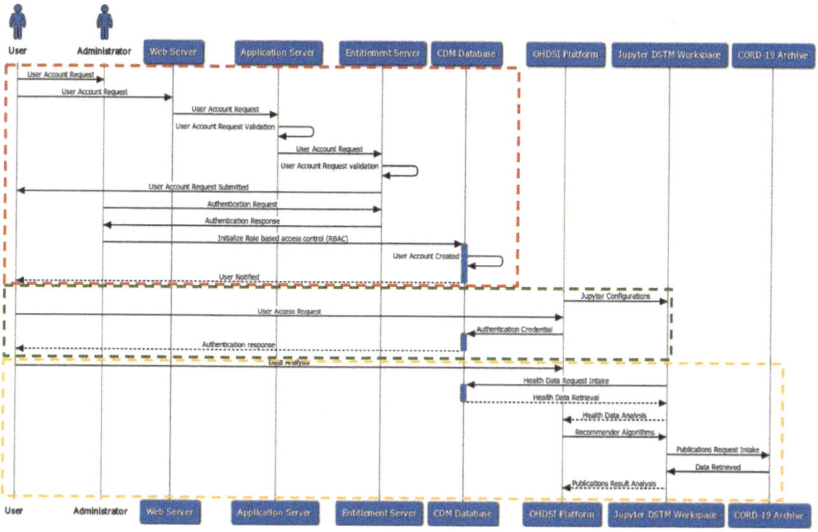

Fig. 7 Sequence diagram of the OnTimeEvidence steps to allow secure access to the analytics workspaces.

interface can be configured to be more secure by having a user community domain in the AWS certificate manager (ACM) service.

4.1 Secure access

The OnTimeEvidence web interface-related files are hosted on a S3 bucket with security options to allow exclusive subdomains accessibility. Traffic can be routed from our custom domain to this S3 bucket on a SSL channel to securely transfer user data and facilitate user communications. The filled user data in the forms are stored in an user entitlement database server (built with MongoDB). The entitlement dB is on a three-cluster sharded server hosted on AWS for high availability, and has services available such as field level encryption to encrypt the data at rest. Sharding is also useful for improving responsiveness and capacity of the entitlement dB. *Nodejs* is used in the application server that connects the OnTimeEvidence web interface (created using *Angularjs*) to the entitlement dB. The database credentials are hidden on the application server and are not accessible to nonauthorized users. Further, we use the *bcryptjs* library to encrypt the administrator's password so that the password is secured even at rest in the entitlement dB, i.e., a third party having access to the encrypted administrator password will not be able to access the data. We connect to the CDM on AWS Redshift through our *nodejs* powered application server. Such a setup allows us to run database query commands on the CDM model to create users with requested credentials and then assign them into certain groups (associated with their roles) for access on certain CDM schema elements in the entitlement dB (Figs. 8 and 9).

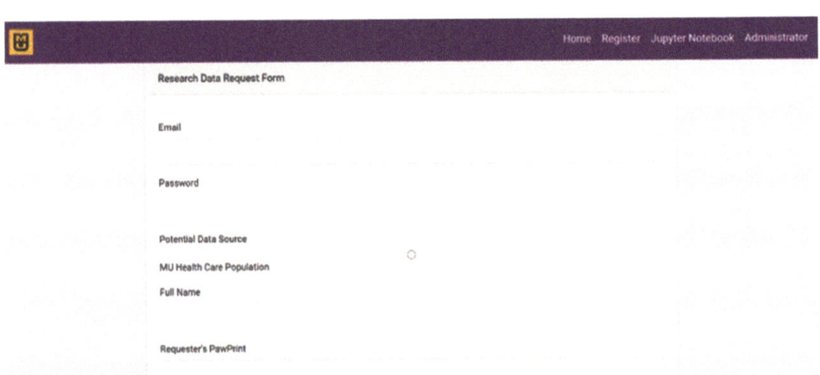

Fig. 8 Administrator web interface to authenticate, authorize, and manage users in the data processing pipeline.

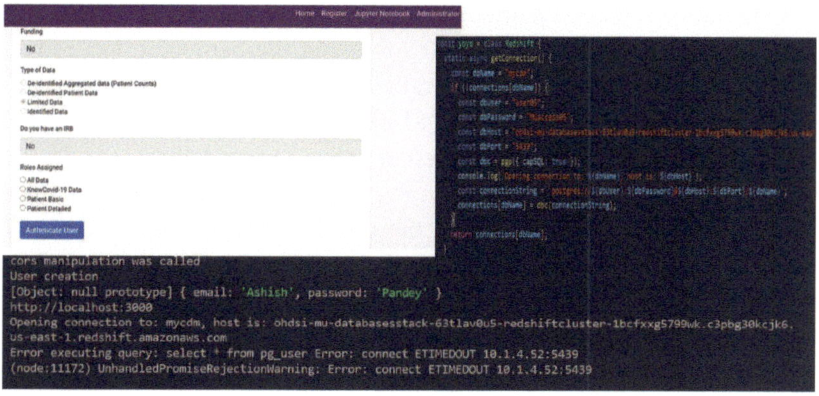

Fig. 9 Role assignment process for restricting and/or allowing users to access functionalities of the data processing pipeline.

4.2 System login and data request

Once the users are provisioned with the proper access roles, a researcher can access the OHDSI environment with the assigned OnTimeEvidence platform credentials. By default, the user will be taken to the JupyterLab environment where data request forms are available (as shown in Fig. 10) to

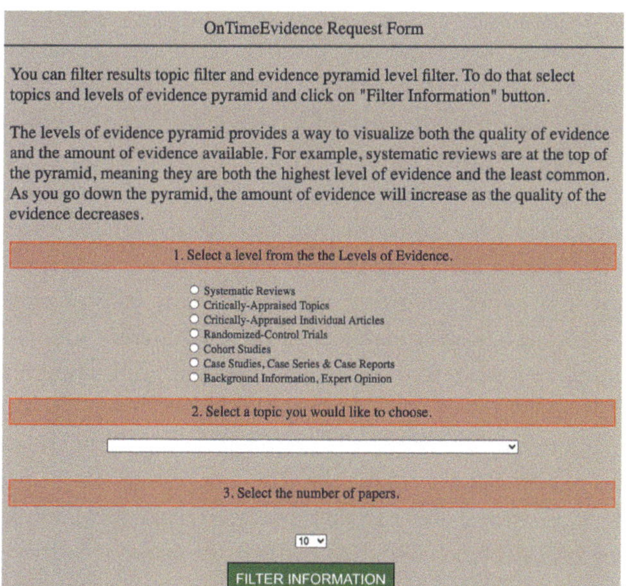

Fig. 10 Data access interface form integrated in the JupyterLab workspace for healthcare data consumers.

allow the submission of a particular CORD-19 request or SynPUF healthcare data request. Based on the user's role and the particular data request elements submitted by the researcher, the access management module, integrated as an extension to the OHDSI platform in OnTimeEvidence, will evaluate if the data elements requested are authorized for access by the related role's permissions. Accordingly, it will either authorize or deny the request. Upon authorization of the data request, the user will be able to retrieve and utilize the data within a Jupyter Notebook and employ the analytic tools available in the workspace using the two configuration modes described below.

4.3 OnTimeEvidence analytics workspace

Once the health data request is received, the platform provides the health consumer Jupyter workspace where the user can conduct an analysis over the SynPUF data stored in the CDM or the CORD-19 dataset stored on an Amazon Redshift cluster. In the following, we detail the process in which researchers can configure their workspace analysis through SynPUF data and DSTM embedded in the Jupyter Notebook for conducting a real-time analysis over publication archives.

In the first configuration mode, the researcher uses a Jupyter Notebook to retrieve and analyze patients' data from the SynPUF dataset. Such a configuration allows researchers to fill a data request, define population selection criteria elements (i.e., patient's year of birth, gender, race, diagnosed conditions, drug treatments), and the data domain elements to be extracted (conditions, drugs, procedures) for the required dataset. Upon submission, the request is processed by a customized JavaScript embedded in the request form, and the data elements included in the request are interpreted and a SQL statement is generated based on the standard schema of the CDM. The SQL statement is then embedded into a new Jupyter Notebook and this notebook instance is made available to the user within the JupyterLab environment. The user can subsequently use the Jupyter Notebook to execute Python code that runs the SQL statement against the CDM repository and retrieves the related dataset. With the data available in the Jupyter Notebook, users can use the analytic Python libraries to run data analyses. Results from this process can be saved in the Jupyter Notebook for further analysis or can be shared with collaborators. Further, the Jupyter Notebook can also be used in conjunction with results generated by the CORD-19 configuration described below to find correlations and gain insights.

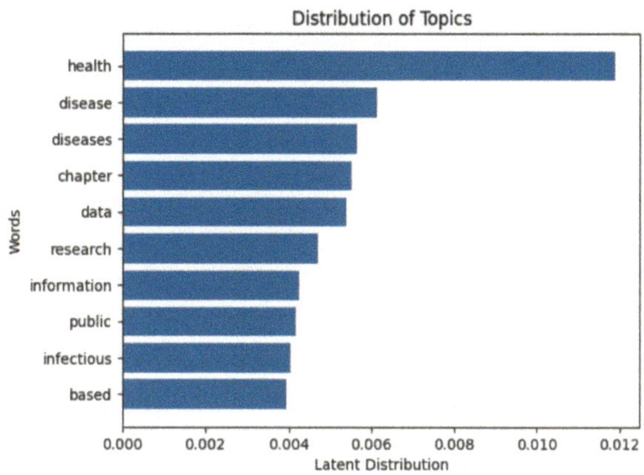

Fig. 11 Distribution chart of topics generated by the DSTM tool from the CORD-19 dataset; this chart is part of the visualizations presented in the analytics workspace.

The second configuration mode includes integrating the DSTM in the Jupyter Notebook to conduct an experiment over the CORD-19 data. Such a configuration allows health-care data consumers to filter their queries according to the Levels of Evidence Pyramid structure to obtain high-quality information from publication archives. The OnTimeEvidence data request form for COVID-19 allows users to select a level (e.g., background information to systematic reviews and meta analyses) and choose a topic from the DSTM that generates a Dirichlet distribution of words within each latent topic that was observed. Once the topic and level of choice are selected by data consumers in the request form, they are used as query parameters on the Jupyter Notebook. Fig. 11 shows a distribution of words for a given topic along with the frequency of that word in the CORD-19 dataset. Health-care data consumers can thus leverage this information to find the latest trends among topics that are pertinent to their research tasks related to the pandemic response.

5 Conclusion—What we have learnt?

Cloud-based platforms are critical for sharing and analyzing the rapidly increasing COVID-19 datasets in a scalable, standard, and secure manner, while also utilizing AI-based tools to automate the data pipeline processing for health-care data consumers (e.g., immunologists, clinical researchers). Our proposed OnTimeEvidence is an exemplar and leverages the OHDSI on the AWS environment to provide users with a scalable platform integrated with a standards-compliant data repository integrated with AI-based data analytics tools. To comply with the privacy requirements for health-care data, we adopted a role-based access control and authorization implementation to define and limit the access to the proper level of data for each user. To expedite the data request processing, OnTimeEvidence includes a data request form that guides users to select the data domain and data identifier elements to retrieve a particular dataset from the OHDSI repository required for a research task. To reduce the randomness found in existing approaches used to extract relevant information, we integrated the DSTM tool that uses the Gibbs sampling algorithm internally to generate a reliable set of results related to COVID-19 publication analytics. Consequently, OnTimeEvidence helps users to discover relationships, e.g., between drugs and genes within a large text corpus of medical research journals. Further, our OnTimeEvidence helps users to customize Jupyter workspaces included in the OHDSI deployment to perform COVID-19-related data retrieval and drill-down analytics to rapidly respond to the pandemic response issues.

References

Abdel-Basset, M., Chang, V., Hawash, H., Chakrabortty, R. K., & Ryan, M. (2021). Fss-2019-ncov: A deep learning architecture for semi-supervised few-shot segmentation of covid-19 infection. *Knowledge-Based Systems*, *212*, 106647.

Abdel-Basset, M., Chang, V., & Mohamed, R. (2020). Hsma_woa: A hybrid novel slime mould algorithm with whale optimization algorithm for tackling the image segmentation problem of chest x-ray images. *Applied Soft Computing*, *95*, 1066423.

Abdel-Basset, M., Chang, V., & Nabeeh, N. A. (2020). An intelligent framework using disruptive technologies for covid-19 analysis. *Technological Forecasting and Social Change*, *163*, 120431.

Ashraf, M. U., Hannan, A., Cheema, S. M., Ali, Z., Alofi, A., et al. (2020). Detection and tracking contagion using IoT-edge technologies: Confronting COVID-19 pandemic. In *2020 international conference on electrical, communication, and computer engineering (ICECCE)* (pp. 1–6). IEEE.

Barik, R. K., Dubey, H., & Mankodiya, K. (2017). Soa-fog: Secure service-oriented edge computing architecture for smart health big data analytics. In *2017 IEEE global conference on signal and information processing (GlobalSIP)* (pp. 477–481). IEEE.

Bayat, A., Szul, P., O'Brien, A. R., Dunne, R., Hosking, B., Jain, Y., et al. (2020). Variant-spark: Cloud-based machine learning for association study of complex phenotype and large-scale genomic data. *GigaScience*, *9*, giaa077.

Blei, D. M., Ng, A. Y., & Jordan, M. I. (2003). Latent Dirichlet allocation. *Journal of Machine Learning Research*, *3*, 993–1022.

Borton, J., Yu, A., Crego, A., Singh, A., Davern, M., & Hair, E. (2010). Data entrepreneurs' synthetic puf: A working puf as an alternative to traditional synthetic and non-synthetic pufs. In *JSM proceedings, survey research methods section*.

Cohen, A., & Nissim, N. (2018). Trusted detection of ransomware in a private cloud using machine learning methods leveraging meta-features from volatile memory. *Expert Systems with Applications, 102*, 158–178.

Dai, B., Ding, S., Wahba, G., et al. (2013). Multivariate Bernoulli distribution. *Bernoulli, 19*, 1465–1483.

Dinakarrao, S. M. P., Sayadi, H., Makrani, H. M., Nowzari, C., Rafatirad, S., & Homayoun, H. (2019). Lightweight node-level malware detection and network-level malware confinement in IoT networks. In *2019 design, automation & test in Europe conference & exhibition (DATE)* (pp. 776–781). IEEE.

Ekin Eren, M., Solovyev, N., Raff, E., Nicholas, C., & Johnson, B. (2020). *COVID-19 Kaggle literature organization.* arXiv e-prints, arXiv–2008.

Friedman, A. A., Letai, A., Fisher, D. E., & Flaherty, K. T. (2015). Precision medicine for cancer with next-generation functional diagnostics. *Nature Reviews Cancer, 15*, 747–756.

García, Á. L., De Lucas, J. M., Antonacci, M., Zu Castell, W., David, M., Hardt, M., et al. (2020). A cloud-based framework for machine learning workloads and applications. *IEEE Access, 8*, 18681–18692.

Griffiths, T. L., & Steyvers, M. (2004). Finding scientific topics. *Proceedings of the National Academy of Sciences of the United States of America, 101*, 5228–5235.

Hörandner, F., Krenn, S., Migliavacca, A., Thiemer, F., & Zwattendorfer, B. (2016). Credential: A framework for privacy-preserving cloud-based data sharing. In *2016 11th international conference on availability, reliability and security (ARES)* (pp. 742–749). IEEE.

Hossain, M. S., Muhammad, G., & Guizani, N. (2020). Explainable AI and mass surveillance system-based healthcare framework to combat COVID-I9 like pandemics. *IEEE Network, 34*, 126–132.

Hripcsak, G., Duke, J. D., Shah, N. H., Reich, C. G., Huser, V., Schuemie, M. J., et al. (2015). Observational health data sciences and informatics (OHDSI): Opportunities for observational researchers. *Studies in Health Technology and Informatics, 216*, 574.

Ioannidis, J. P., Salholz-Hillel, M., Boyack, K. W., & Baas, J. (2020). *The rapid, massive infection of the scientific literature and authors by covid-19.* bioRxiv.

Kricka, L. J., Polevikov, S., Park, J. Y., Fortina, P., Bernardini, S., Satchkov, D., et al. (2020). Artificial intelligence-powered search tools and resources in the fight against covid-19. *EJIFCC, 31*, 106.

Makadia, R., & Ryan, P. B. (2014). Transforming the premier perspective® hospital database into the observational medical outcomes partnership (OMOP) common data model. *Egems, 2*(1), 1110.

Matos, D. R., Pardal, M. L., Adao, P., Silva, A. R., & Correia, M. (2018). Securing electronic health records in the cloud. In *Proceedings of the 1st workshop on privacy by design in distributed systems* (pp. 1–6).

Milken Institute. (2020). *COVID-19 treatment and vaccine tracker.* URL: https://covid-19tracker.milkeninstitute.org.

Murad, M. H., Asi, N., Alsawas, M., & Alahdab, F. (2016). New evidence pyramid. *BMJ Evidence-Based Medicine, 21*, 125–127.

Otoom, M., Otoum, N., Alzubaidi, M. A., Etoom, Y., & Banihani, R. (2020). An IoT-based framework for early identification and monitoring of covid-19 cases. *Biomedical Signal Processing and Control, 62*, 102149.

Pickett, B. E., Sadat, E. L., Zhang, Y., Noronha, J. M., Squires, R. B., Hunt, V., et al. (2012). ViPR: An open bioinformatics database and analysis resource for virology research. *Nucleic Acids Research, 40*, D593–D598.

Sackett, D. L. (1997). Evidence-based medicine. In *Seminars in perinatology* (pp. 3–5). Elsevier.

Sharma, S., Chen, K., & Sheth, A. (2018). Toward practical privacy-preserving analytics for IoT and cloud-based healthcare systems. *IEEE Internet Computing, 22*, 42–51.

Simmhan, Y., Aman, S., Kumbhare, A., Liu, R., Stevens, S., Zhou, Q., et al. (2013). Cloud-based software platform for big data analytics in smart grids. *Computing in Science & Engineering, 15*, 38–47.

Tuli, S., Tuli, S., Tuli, R., & Gill, S. S. (2020). Predicting the growth and trend of covid-19 pandemic using machine learning and cloud computing. *Internet of Things, 11*, 100222.

Wiggins, J. (2018). *Create data science environments on AWS for health analysis using OHDSI.* URL: https://aws.amazon.com/blogs/big-data/creating-data-science-environments-on-aws-for-health-analysis-using-ohdsi/.

Yue, X., Wang, H., Jin, D., Li, M., & Jiang, W. (2016). Healthcare data gateways: Found healthcare intelligence on blockchain with novel privacy risk control. *Journal of Medical Systems, 40*, 1–8.

Zhang, Y., Calyam, P., Joshi, T., Nair, S., & Xu, D. (2018). Domain-specific topic model for knowledge discovery through conversational agents in data intensive scientific communities. In *2018 IEEE international conference on big data (big data)* (pp. 4886–4895). IEEE.

CHAPTER 7

Threat model and security analysis of video conferencing systems as a communication paradigm during the COVID-19 pandemic

Raiful Hasan and Ragib Hasan
Department of Computer Science, University of Alabama at Birmingham, Birmingham, AL, United States

1 Introduction

Video conferencing is a live, visual connection between two or more parties regardless of their locations. It is modern technology that has changed our communication paradigm. This technology is widely used in every sector including business, academics, healthcare, social communication, etc. The demands of video conferencing and collaboration tools have increased excessively, especially during the COVID-19 pandemic. All academic institutes, regardless of their size, in the United States and some other countries such as Canada, Australia, the United Kingdom, etc., have moved online during the pandemic. Besides, people are spending more time with friends and families during the lockdown in video conferencing tools. The casual purpose of video conferencing is called a video call or a video chat. Experts predict that the trend will continue in the future, even after the end of the COVID-19 pandemic (Chen et al., 2020; Fatani, 2020). It will not only change our fundamental communication concept, but also cause a permanent shift in how people work. Activity will increase in those video conferencing tools; more activity means more bad actors will look for vulnerabilities and other ways to exploit the system.

The security of video conferencing is becoming a growing issue for companies and various government agencies. The main concern is the fear of unauthorized access to private and confidential data. For that, many Fortune 500 companies have adopted a cautious approach when using video conferencing with their employees. The video conferencing sessions are transmitted using the Internet Protocol (IP) or Integrated Services Digital Network (ISDN) standard (Bocker, 2012; Simpson, 2013). In the IP standard, the

Novel AI and Data Science Advancements for Sustainability in the Era of COVID-19
https://doi.org/10.1016/B978-0-323-90054-6.00009-X

video packets travel over the transmission medium. Intruders can install sniffing tools in the network path (e.g., at the Switch) to read these packets if there is no secure encryption. In ISDN, the transmitted packets switched their formats multiple times throughout the journey from source to destination. Also, the path of the information is uncertain, so this standard cannot provide maximum privacy. Attackers can use some off-the-shelf tools to monitor these packets, such as ISDN line tester tools.

Another security concern of video conferencing technology is the physical location of the provider's infrastructure. It is important in international and cross-country video conferencing. The ownership of the provider company, the relationship with the government, and the local law also influence security and privacy. Many countries controlled the telecommunication provider and ruled them to have access to the data or intercept the communication channel (Whittaker, 2013; Yang & Xu, 2018). Besides, there are also vulnerabilities in video conferencing equipment, such as network instruments and communication devices. Rival countries often use these instruments to spy on other countries. Recently, the United States passed a bill to replace the equipment made by Huawei[a] and ZTE[b] for security concerns (Drezner, 2020).

As a result, the security, privacy, and vulnerabilities of video conferencing tools are an active research area and have gained attention in both academics and industries. There should be a plan to consider all possible threats against video conferencing; otherwise, the system would be vulnerable to cyber-physical attacks. Proper threat modeling helps to analyze the security breaches and system requirements (Mellado, Blanco, Sánchez, & Fernández-Medina, 2010; Myagmar, Lee, & Yurcik, 2005; Scandariato, Wuyts, & Joosen, 2015). Adaptation of threat modeling is important before implementing the video conferencing system because it helps to understand the nature of threats as well as improve the system performance. Threat modeling helps to understand the complexity of the system and identify the possible dangers. The model analyzes all identified threats based on their priority and helps to decide whether it mitigates the threats or accepts the risk.

Contributions: The key contributions in this paper are as follows:

1. A complete threat analysis is provided by characterizing threat components (e.g., assets, potential attackers, etc.).

[a] Huawei—https://www.huawei.com/us/.
[b] ZTE—https://www.zte.com.cn/global/.

2. The STRIDE threat model is adopted to identify potential attacks and vulnerabilities.
3. We have suggested a mitigation plan to prevent potential threats and vulnerabilities.

Organization: The rest of the paper is organized as follows: Section 2 provides the background of threat modeling and video conferencing technology. Sections 3 and 4 identify the assets of video conferencing tools and entry points of attackers, respectively. Section 5 presents the attacker model. Sections 6 and 7 describe the vulnerabilities and mitigation strategies. The conclusions and future work are presented in Section 8.

2 Background

2.1 Privacy and security in COVID-19

User privacy and data protection have brought concerns during the pandemic. Personal and social programs adopted on video conferencing systems were unprecedented even a few years ago including virtual birthdays, funerals, virtual therapies, etc. Millions of users globally had a similar type of experience when the school went remote, with work and business meetings done on virtual. Actually, people shifted their entire lifestyle online and barely regulated virtual sessions were experiencing cyberattacks, racism, and pornography. Experts suggest that data security and user privacy could be threatened at this time for spending more time online and remote work nature (Uberti, 2020). Users are giving tons of personal data to get things they need. Most importantly, people know very little about the current privacy laws or regulations. One study found that more than 63% of Americans know very little or nothing about their data privacy (Uberti, 2020).

2.2 Threat modeling

Threat modeling is a process to understand system security and identify the potential vulnerabilities with a cost-effective mitigation technique for the attacks (Shostack, 2014). This technique is widely used in multiple sectors, including software, applications, distributed systems, business process, etc. Due to the emergence of new technology and tools, it is quite impossible to implement a fully secure system. On the other hand, building and maintaining a secure system is expensive. Authorities have to decide the priority and identify the assets and maintain a balance between security and cost. The priority is determined by the value and potential risk of it. The threat model consists of five components: (i) *Identifying Assets*, (ii) *Defining the Trust Levels of Users*, (iii) *Attackers Model*, (iv) *Identifying Threats*, and (v) *Mitigation Plan*.

2.3 Video conferencing technology

A video conferencing system consists of the following components:

- Endpoints—personal computer, smartphone, telepresence system, etc.
- Peripheral equipment—webcam, headset, projector, etc.
- Video conferencing server—to process the video conference.
- Software infrastructure—content transmission, storage, integration, etc.

There are two video conferencing modes based on the participants: (i) Point-to-Point Video Session and (ii) Multipoint Video Conferencing. A point-to-point video session connects two different users or locations. Multipoint or group video conferencing connects more than two users or sites. A multipoint conferencing unit (MCU) is needed to tie the locations together in the multipoint conferencing mode. The MCU can be embedded on a remote server in the system. The MCU-supported system is called centralized multipoint video conferencing. Some systems are capable of conducting multipoint conferencing without any support from the MCU. This approach can afford better quality due to the absence of any central repository and greater convenience (participants can make ad hoc multipoint calls regardless of MCU availability). This technique is called decentralized multipoint video conferencing. However, it requires increased network bandwidth since every node transmits to every other node directly. Fig. 1 shows the different existing video conferencing techniques.

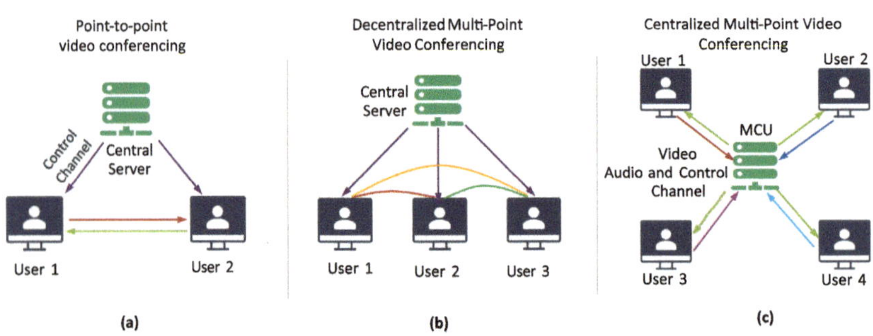

Fig. 1 Different video conferencing modes (A) point-to-point video conferencing, (B) decentralized multipoint video conferencing, and (C) centralized multipoint video conferencing.

2.4 Growth of video conferencing tools in the COVID-19 era

The COVID-19 pandemic has begun a cultural shift worldwide. The situation has pushed people to stay at home and work remotely. Social distancing, lockdown, and work-from-home policies have moved the regular in-person activity online. Video conferencing apps allow us to keep in touch with our dearest ones and conduct our day-to-day lives in education, business, health, etc. Some countries have even conducted their cabinet meeting via video conferencing during this pandemic, which was uncommon. The COVID-19 pandemic has bolstered the growth rate of these tools. According to Statista,[c] the video conferencing app was downloaded nearly 27 million times in March 2020; the number was 2.1 million in January 2020 (Richter, 2020). Besides, the statistics reported that both the daily active users and meeting participants have increased during that time. For example, Zoom (Zoom Video Communications Inc.) active users increased from 10 million to 200 million in just three months (REUTERS, 2020), whereas Teams (Microsoft) reported that it had surpassed 75 million daily active users (Zaveri, 2020). Over 90,000 schools from 20 countries and 96% of the top US universities used Zoom during the COVID-19 outbreak (Yuan, 2020). However, some governments and organizations banned some third-party video conferencing applications due to security issues. For example, SpaceX[d] and NASA[e] banned Zoom; Standard Chartered[f] has banned both Zoom and Google Hangouts (Jain, 2020).

3 Assets of video conferencing tools

An asset of any system is something valuable or a tangible or intangible entity. Assets are the primary target of the system, and the attackers want to gain access, modify, or destroy. Identifying and measuring the asset is the primary step of threat modeling. The asset must have the business value that should be protected from the attacker. Table 1 shows the asset with the description of the video conferencing tools.

Fig. 2 shows a generic architecture and set of assets of video conferencing tools.

[c] Statista—https://www.statista.com.
[d] SpaceX—https://www.spacex.com.
[e] NASA—https://www.nasa.gov.
[f] Standard Chartered—https://www.sc.com/en/.

Table 1 Asset and description of the video conferencing tool.

ID	Asset name	Description
A1	End user	The asset on the user side.
A1.1	User credentials	User login credentials authenticate a user when logging into web/ desktop/mobile version.
A1.2	User devices	An end-user device is a personal computer (desktop or laptop), consumer device (smartphone, tablet)
A1.3	Device credentials	End-user device information, such as smartphone IMEI, identification key, etc.
A1.4	Application	An application is a program or group of programs that is designed for the end-user. It is capable of dealing with user inputs (video/audio) and provides the appropriate outputs.
A1.5	Data in application	Sometimes end application stores data related to the users in the local devices. Intruders could grab these data if the device compromised.
A1.6	Peripheral equipment	Peripheral devices add functionality but not essential every time. For example, webcam, microphone, headset, etc.
A2	Conference server	The asset on the server side.
A2.1	Data storage	The data center facility for the purpose of storing, processing, and provide the data.
A2.2	Conference session	Identification of the conference session, such as link, numbers, key, etc.
A2.3	Video, audio data	Contents of the video conference session. However, it can be open if the user makes it public.
A2.4	Log file	Participant's location, device configuration, time, participant's number, etc.
A3	Infrastructure	Physical and organizational facilities
A3.1	Service server	Responsible for video processing, content transmission, and integration.
A3.2	Software support	The internal program that is not directly in the application but gives support to run the systems.

4 Point of entry

Attacks can come from anywhere. It could be internal or external. Attackers target the network, employee software, email, unsecured WiFi, website, even the staff's mobile to enter the system. Entry points are referred to as target areas where the attackers are able to exploit and gain access. With technological improvement, the points of entry change over time. For example, IoT and smart sensors create a new attack surface, which was

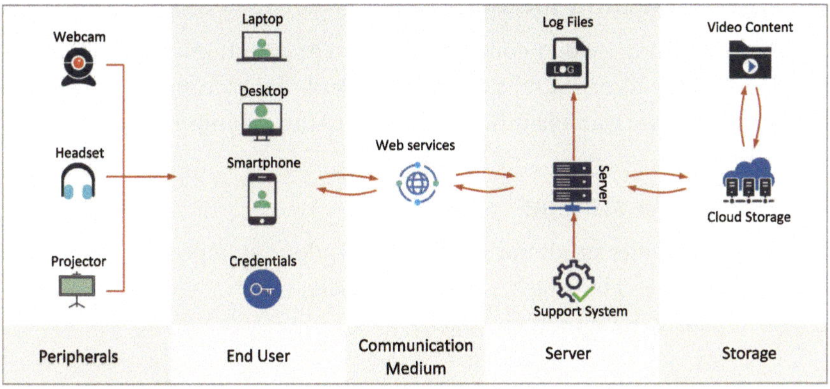

Fig. 2 A simplified architecture and set of assets of video conferencing tools.

not common in the previous decade. Here are the entry points of the video conferencing system.

4.1 Peripheral equipment

Nowadays, smart peripherals such as smart headphones, webcams, etc., are becoming more common in video conferencing. Most of these devices are not secure from cyberattacks (Nissim, Yahalom, & Elovici, 2017; Vlajic & Zhou, 2018). The data communication channel between these devices and the host devices is vulnerable. Besides, Cybercriminals can use these devices as access points to get into the network.

4.2 Device or application

To attack the participant's device is the most common way to enter a conference session. Because user devices can be compromised in various ways, the attacker can disrupt the conference call, steal sensitive information such as credentials, call records, etc., and access the private session by malware or phishing attacks.

4.3 Conference session

The attackers get into a video conference and make unwanted comments, threatening others, interjecting racism, sharing the malicious link, or making the video unavailable to others. Usually, new and nontechnical users face this type of situation. Sometimes, the attacker coordinates a group attack with a specific session. Though this is not a security flaw, the service provider must take action for the uninterrupted video session.

4.4 Communication medium

The participants of a video conferencing system communicate through the internet with an IP or ISDN standard protocol. Eavesdroppers can exploit these protocols and gain unauthorized access to the communication channel.

4.5 Cloud data storage

The service provider stores the call data in the cloud storage and periodically sends the updates. The attacker could get inside the data center both physically or virtually to launch an attack.

4.6 Application system server

The application server is the heart of the system, and the attacker can launch an attack on it. The server is the most crucial entry point of the conferencing system because previous entries and subsequent attacks only hamper the participants. However, if the attacker gets into the server, then the whole system will be compromised.

5 Attack model

The attack model presents the information that is used to think like an attacker. This section will characterize the user's accessibility to the system and the attacker's motives and capabilities.

5.1 Users or potential attacker

Identifying the system users and defining their role is crucial in threat modeling because threats primarily originate from two sources: internal users—who have access authorization, and external agents—who do not have access authorization. Based on accessibility, there are three types of users in video conferencing systems.

5.1.1 Participants

Participants are those who take part in video conferencing and end-users of the system. The host, users with credentials, and users without credentials are the participants. They have access to their own devices and can be an attacker who has a related background, or sometimes a group of participants can organize an attack.

5.1.2 Organization administrator

Organization administrators control all users or part of user's accessibility of the organization. They can monitor a call within an organization and could access the conference content in the cloud.

5.1.3 System administrator

System admins are responsible for system operation, maintenance, and management. Usually, a group of people do this job and have the highest level of authorization to ensure the correct operation of software and hardware. The super admin, the developer of the system, the database administrator, and other support administrators have admin accessibility. Generally, administrators are very trustworthy; they are highly technical people to attack any part or the whole system.

5.2 Trust levels of system users

Trust levels represent the security rules to the entities of the system. Internal agents are the most trustworthy; however, they are more vulnerable and challenging than external agents because insiders are authorized to access the resources. For example, a super admin has the highest level of privilege to all components of the system; and also has access to all physical entities. If they became an attacker, it would be difficult to prevent the risk. Table 2 shows the trust level of the users.

Table 2 Trust levels of the users of video conferencing system.

ID	User	Description
TL1	Participants	The lowest level of trust; end users and hosts.
TL1.1	Host	Who initiates the conference call and invites the other participants, has access to *A1.2* and *A1.4*.
TL1.2	With credentials	Similar access level of the host, but instead of initiates a new call, they join a call.
TL1.3	Without credentials	Do not have access to *A1.4*; sometimes can join an invited call.
TL2	Organization administrator	Trust level between an individual user and system admin
TL2.1	Organization IT personnel	They have authorization for all members of the organization. They can access to *A2.3* as well as *A1.1*, *A1.4* for individual users.
TL3	System administrator	Highest level of trust

Continued

Table 2 Trust levels of the users of video conferencing system—cont'd

ID	User	Description
TL3.1	Developer	Developers are responsible for creating the applications or systems that run on the computer or another device. Usually, they have access to *A3.1* and *A3.2*. Some developers have *T3.4* trust levels.
TL3.2	DBA admin	Database administration (DBA) creates and maintains databases. DBA is a critically important role as they are responsible for database updates, storage, security, and troubleshooting. DBA has authorization on *A2.1*, *A2.2*, and *A2.3*.
TL3.3	Support admin	The support admin is responsible for technical issues relating to software implementation, function, and upgrades. They can access *A3.1* and *A3.2*. Sometimes they have superadmin access.
TL3.4	Superadmin	They are responsible for the whole system operation and maintenance. Superadmin has access to all components of the service and system.

5.3 Attacker's motives

The motivation and goals of attackers depend on the individuals and systems. The primary motivation behind the attack is to access or control any part of the system or assets. Some potential motives are as follows.

5.3.1 Financial gain

Financial gain is the most common motivation behind any attack. According to a study by Verizon,[g] more than 60% of attacks were motivated by financial gain in 2020 (Verizon, 2020). These attacks were launched by private criminals or hired by the rogue nation.

5.3.2 Espionage

Two types of data are targeted for espionage—political movement and national interest. For example, in the COVID-19 era, some renowned research institutions have been attacked by different state-owned groups from other countries (Mehrotra, 2020; Orme, 2020). The attacker's target was to steal Coronavirus research data.

[g] Verizon—https://www.verizon.com/.

5.3.3 Business rival

Attackers look for secrets that they or their employers can use to dominate in the market. Usually, the same type of organization tries to find security breaches from other companies for a market share. For example, ThyssenKrupp AG[h] (a German steel company) discovered that their trade secret had been compromised by hackers from southeast Asia in 2016 (Auchard & Käckenhoff, 2020).

5.3.4 FIG (Fun, Ideology, Grudge)

The FIG category consists of several other motivations behind the attack, including Fun, Ideology, and Grudge. The most usual tactic of these attacks is to steal credentials and valuable data from the systems. Attackers use web applications, LAN access, or physical access to conduct the attack.

6 Threats and vulnerabilities

6.1 STRIDE threat modeling process

STRIDE is an acronym that stands for six categories of security risks. The STRIDE model is proposed by Microsoft to identify the threats of a system (Shostack, 2014). STRIDE stands for:

- Spoofing—access with a fake identity
- Tampering—modification of data or code
- Repudiation—denying the signature authenticity
- Information disclosure—expose the data to an unauthorized person
- Denial of service—making a service unavailable
- Elevation of privilege—gain control with improper authorization

The STRIDE threat model is developed from a developer's perspective. While designing security for applications, developers and software engineers must consider security threats early in the Software Development Life Cycle. Table 3 shows the different types of threats and vulnerabilities according to STRIDE threat modeling.

6.2 More attacks and vulnerabilities

In recent years, security experts have been warning about the growing risk and potential impact of cyber-physical attacks on the system. The list of attacks and techniques is increasing rapidly due to technological advancement. The video conferencing system is highly dependent on the physical

[h] ThyssenKrupp—https://www.thyssenkrupp.com.

Table 3 Threats of a video conferencing system in STRIDE threat modeling.

STRIDE	ID	Potential threats
Spoofing	T1.1	**Participants Identity Theft:** The attacker can steal the credentials from the participants and then use that identity to accomplish a goal. The attacker does not need any special resources for this attack.
	T1.2	**Source Device Spoofing:** The attacker can steal the user's device data and pretend to be a legitimate device by performing the authentication to the server.
	T1.3	**Content Spoofing:** An adversary modifies the video/audio content between participants. Any data can be spoofed, including text messages, files, or communication protocols.
	T1.4	**Action Spoofing:** The attacker gets inside the system and disguises an action for another one. The user performs this action as a real action and gets attacked.
Tampering	T2.1	**Call Content Tampering:** Conference call transfer through the Internet in data packets, an intruder can intercept these packets and modify them.
	T2.2	**Cloud Storage Tampering:** The attacker may insert, delete, or manipulate data to the storage. Tampering through the Application Programming Interfaces (APIs) can affect the availability and security of the cloud storage.
	T2.3	**Log Files Tampering:** Logs tampering attacks involve an attacker gain unauthorized access to the system and injecting, deleting, or otherwise tampering the files. They do that to masking for other malicious activities.
Repudiation	T3.1	**Content Repudiation:** The participants deny or claim not receiving, sending, or editing the data. The attacker can change the authoring information by the malicious program and log the wrong information.
	T3.2	**Activity Hiding:** The intruders hide the activity after getting inside the system. They try to deny for being convicted in future investigations.
Information disclosure	T4.1	**Participant Information Disclosure:** The attacker may unauthorizedly access the participant's information such as name, age, location, etc. Later they can use this data or publish publicly.
	T4.2	**Device Information Disclosure:** Device information, such as device location, device identity, device type, etc. could disclose or can use for further attack.
	T4.3	**Call Data Disclosure:** The attacker may copy or transmit the data to another unauthorized source, without affecting the original data during the video conferencing.

Table 3 Threats of a video conferencing system in STRIDE threat modeling—cont'd

STRIDE	ID	Potential threats
	T4.4	**Application Error Disclosure:** The error or warning messages of the application may expose the crucial and sensitive information of the internal function and architecture. The message may contain the location of the file, cause of the errors, internal calculations, etc.
DOS	T5.1	**Denial of Service (DoS):** To make the conferencing unavailable to legitimate users. The attacker accomplishes this by flooding the system with traffic or sending it information that triggers a crash. DoS attack is performed by a single computer to launch the attack.
	T5.2	**Distributed Denial of Service (DDoS):** Multiple sources combinedly attack the conferencing system; it is a type of DoS attack. Even the participant's device can use as a source of an attack.
Elevation of privilege	T6.1	**Unauthorized Admin Privilege:** The attacker tries to obtain crucial privileges by a variety of techniques, such as phishing, brute force, etc. The legitimate owner might not even realize the account has been compromised.
	T6.2	**Unauthorized Access:** The superadmin, system admin, and the DBA gain unauthorized access to the participant's call content, credentials, information, etc.
	T6.3	**Privilege Abuse:** The attacker exploits some features reserved for the privileged or administrator user; they make them accessible to those who are not authorized to use them.

server and storage; the telemetry vulnerabilities of the hardware can allow cyber tampering and the interception of data. Table 4 shows some more threats that are not listed in Table 3.

7 Mitigation strategies

Several studies have been conducted and proposed to implement a tamper-proof system to protect privacy and security (Chang, Muñoz, & Ramachandran, 2020; Chauhan, Kaur, & Chang, 2020; Correia, Liu, & Xu, 2020; Suciu, Stefanescu, Beceanu, & Ceaparu, 2020). Each threat category has a corresponding set of countermeasure techniques that should be used to reduce risk. Though the appropriate countermeasure depends on the specific attack, we have summarized the mitigation plan in Table 5 for the video conferencing tools.

Table 4 Threats of a video conferencing system other than STRIDE model.

Attack	ID	Potential threats
Physical	T7.1	**Hardware Integrity Attack:** The intruders can exploit the system or install a malicious part on hardware to espionage the conference session (Villasenor, 2011). Most of these attacks are related to firmware. Firmware is a software that comes with the hardware and is stored in ROM or flash memory.
	T7.2	**Bypassing Physical Security:** System facilities often used some physical security, such as traditional locks, electronic entry systems, RFID tags for tracking assets, etc. However, these security is not sufficient for the well-planned attacks.
Information engineering	T8.1	**Excavation:** The attacker sends the syntactically invalid or nonstandard request to the system and gathers intelligence about the target by the various responses, such as error response, URL pattern, etc.
	T8.2	**Reverse Engineering:** The attacker can target the customer application and reveal the structure, function, and composition of an object by reverse engineering. Nowadays, intruders can launch an attack easily because many reverse engineering tools are available in the market for commercial purposes.
	T8.3	**Footprinting:** Adversary learns the information of call sessions, including associated ports, networks, peripherals, etc., through footprinting. Though this is not directly harmful, it may often pave the way for more damaging attacks.
Probabilistic	T9.1	**Fuzzing:** Fuzzing is a software security and functionality testing method that discover the weakness in the system from the various input. It is a black box technique and totally free from any preconceptions about the system.
	T9.2	**Brute Force:** The attacker launches an attack on a trial-and-error basis to gain access to valuable assets and secrets of the systems. The attack exhaustively explores all the possible ways to enter the system. The potential secrets include credentials, encryption keys, database lookup keys, etc.

Table 5 Mitigation strategies.

Threat	ID	Mitigation plan
Spoofing	T1.1, T1.2, T1.3, T1.4	**Encryption:** All credentials must be encrypted by a strong encryption algorithm. It needs to make sure that no one has been able to retrieve the actual credentials in any part of the communication, even the system administrator.
		Zero-Trust Approach: The system needs to implement protocols that rely on trust relationships as little as possible. In addition, all connections must be verified through a proper security guideline.
		Packet Filtering: All data are transmitted across the network as the packet. The packet filter restricts the suspicious packet with the conflicting source address. It is useful for IP spoofing attacks.
Tampering	T2.1, T2.2, T2.3	**Authorization:** The strong authorization has to be implemented that will confirm only the authenticated user can perform a given action on a specific resource.
		Data Signing and Hashing: The data signing and hashing will prevent data alteration in communication between the participants and the system.
Repudiation	T3.1, T3.2	**Use Digital Signatures:** In data transfer, the system needs to provide nonrepudiation. Digital signatures create one-way hashes that are similar to the checksum function and prevent repudiation.
		Secure Audit Trails: All activities have to be logged in and recorded for future auditing. This log will assist in access control monitoring and can provide audit trails when investigating an incident.
Information disclosure	T4.1, T4.2, T4.3, T4.4	**Avoiding Self-Signed Certificates:** A trusted certificate authority (CA) has to be used for both external and internal communications.
		Avoid Secrets in Plaintext: There should be a policy to not store, transmit, and display secrets in plaintext at any point of the communication.

Continued

Table 5 Mitigation strategies—cont'd

Threat	ID	Mitigation plan
Denial of service	T5.1, T5.2	**Validate and Filter Input:** All incoming traffic has to be validated and automatically filtered out from the malicious automated bot requests. **Throttling Techniques:** Usually, the video conferencing service provider has a large bandwidth. But to prevent DoS or DDoS attacks, it has to implement throttling techniques. Also, some architecture has to perform, including servers in different data centers, data centers in different networks, etc.
Elevation of privilege	T6.1, T6.2, T6.3	**Principle of Least Privilege (POLP):** A security concept to limiting access if they absolutely do not need that. The concept is not only for the users but also for the hardware, systems, process, applications, etc. **Access Control:** Access control is the application of constraints on whom or what can perform attempted actions or access resources that they have requested. There should be proper system-wide access control. Recent research suggests IoT-based self-adapt access control for medical data (Yang, Zheng, Guo, Liu, & Chang, 2019).
Physical attack	T7.1, T7.2	**Foster a Culture of Security:** To defend from the physical attack and secure the hardware, the providers have to adopt awareness within the entire system and employees. **Tamper-Resistant System Design:** There should be security primitives as well as design and architecture-based countermeasures for hardware tampering and counterfeiting.
Information engineering	T8.1, T8.2	**Proper Error Response:** In order to protect the secret, the system has to give a short and proper error response. **Code Obfuscation Techniques:** This is a mechanism for hiding the original code, algorithm, data structure from the unauthorized reverse engineering process. The production software must be obfuscated.
Probabilistic attack	T9.1, T9.2	**Account and IP Lockouts:** There should be a policy to block the account after multiple unsuccessful login attempts. The system needs to block the device and associated IP address. **Comprehensive Fuzz Testing:** The production application has to be tested thoroughly during the development phase to detect any unexpected behavior and ensure the system behaves as per specification.

8 Conclusion and future work

This paper presents the video conferencing system's security threat modeling and analysis. The security team of the video conferencing system should address these vulnerabilities and threats during the development cycle. A detailed mitigation plan is also presented; they should apply those mitigation approaches. It would be beneficial for the system administrator to monitor and alert the requirements. In addition, this model can be useful for systems designers, researchers, information security practitioners to identify security threats and vulnerabilities. It will help to develop automated testing tools at the early stages of the systems development life cycle for the video conferencing systems. Due to technological improvements, new security threats and attacks will be introduced in the future. Threat modeling will help to understand and classify these upcoming threats and attacks. We will focus on developing a prototype of the video conferencing system and address all the security concerns mentioned in this paper. Besides, we will analyze new threats and vulnerabilities within this system.

Acknowledgments

This research was supported by the National Science Foundation through awards DGE-1723768, ACI-1642078, CNS-1351038, and ECCS-1952090 and by the National Institutes of Health Grant 1R21HD095270-01.

References

Auchard, E., & Käckenhoff, T. (2020). *ThyssenKrupp secrets stolen in 'massive' cyber attack.* https://www.reuters.com/article/us-thyssenkrupp-cyber-idUSKBN13X0VW (Last accessed 05 May 2021).

Bocker, P. (2012). *ISDN the integrated services digital network: Concept, methods, systems.* Springer Science & Business Media.

Chang, V., Muñoz, V. M., & Ramachandran, M. (2020). Emerging applications of Internet of things, big data, security, and complexity: Special issue on collaboration opportunity for IoTBDS and COMPLEXIS. *Computing, 102*, 1301–1304. https://doi.org/10.1007/s00607-020-00811-y.

Chauhan, R., Kaur, H., & Chang, V. (2020). An Optimized Integrated Framework of Big Data Analytics Managing Security and Privacy in Healthcare Data. *Wireless Personal Communications, 117*, 87–108. https://doi.org/10.1007/s11277-020-07040-8.

Chen, T., Peng, L., Jing, B., Wu, C., Yang, J., & Cong, G. (2020). The Impact of the COVID-19 Pandemic on User Experience with Online Education Platforms in China. *Sustainability, 12*, 7329. https://doi.org/10.3390/su12187329.

Correia, A. P., Liu, C., & Xu, F. (2020). Evaluating videoconferencing systems for the quality of the educational experience. *Distance Education, 41*, 429–452. https://doi.org/10.1080/01587919.2020.1821607.

Drezner, D. (2020). *What, exactly, is the Trump administration's 5G policy?*. https://www.washingtonpost.com/outlook/2020/02/10/what-exactly-is-trump-administrations-5g-policy/ (Last accessed 05 May 2021).

Fatani, T. H. (2020). Student satisfaction with videoconferencing teaching quality during the covid-19 pandemic. *BMC Medical Education*, *20*, 1–8. https://doi.org/10.1186/s12909-020-02310-2.

Jain, R. (2020). *Zoom and Google hangouts banned by large companies and government organizations*. https://www.businessinsider.in/tech/news/companies-and-organisations-that-have-banned-zoom-and-google-hangouts/articleshow/75156719.cms (Last accessed 05 May 2021).

Mehrotra, K. (2020). *Hackers target California University leading Covid research*. https://www.bloomberg.com/news/articles/2020-06-04/hackers-target-california-university-leading-covid-19-research (Last accessed 05 May 2021).

Mellado, D., Blanco, C., Sánchez, L. E., & Fernández-Medina, E. (2010). A systematic review of security requirements engineering. *Computer Standards & Interfaces*, *32*(4), 153–165. https://doi.org/10.1016/j.csi.2010.01.006.

Myagmar, S., Lee, A. J., & Yurcik, W. (2005). Threat modeling as a basis for security requirements. In *Symposium on requirements engineering for information security (SREIS)* (pp. 1–8). Citeseer.

Nissim, N., Yahalom, R., & Elovici, Y. (2017). USB-based attacks. *Computers & Security*, *70*, 675–688. https://doi.org/10.1016/j.cose.2017.08.002.

Orme, J. (2020). *Successive cyber attacks shut down European super-computers working on Covid-19 research*. https://techerati.com/news-hub/successive-cyber-attacks-shut-down-european-supercomputers-working-on-covid-19-research/ (Last accessed 21 May 2021).

REUTERS. (2020). *Zoom's daily active users jumped from 10 million to over 200 million in 3 months*. https://venturebeat.com/2020/04/02/zooms-daily-active-users-jumped-from-10-million-to-over-200-million-in-3-months/ (Last accessed 21 May 2021).

Richter, F. (2020). *Video chat apps rise to prominence amid pandemic*. https://www.statista.com/chart/21268/global-downloads-of-video-chat-apps-amid-covid-19-pandemic/ (Last accessed 05 May 2021).

Scandariato, R., Wuyts, K., & Joosen, W. (2015). A descriptive study of Microsoft's threat modeling technique. *Requirements Engineering*, *20*, 163–180. https://doi.org/10.1007/s00766-013-0195-2.

Shostack, A. (2014). *Threat Modeling: Designing for security*. John Wiley & Sons.

Simpson, W. (2013). *Video Over IP: IPTV, Internet Video, H. 264, P2P, Web TV, and Streaming: A complete guide to understanding the technology*. Taylor & Francis.

Suciu, G., Stefanescu, S., Beceanu, C., & Ceaparu, M. (2020). Webrtc role in real-time communication and video conferencing. *1–6.*. Global Internet of Things Summit (GIoTS): IEEE.

Uberti, D. (2020). *Coronavirus cybersecurity fallout might not be felt for weeks or longer*. https://www.wsj.com/articles/coronavirus-cybersecurity-fallout-might-not-be-felt-for-weeks-or-longer-11585128601 (Last accessed 05 May 2021).

Verizon. (2020). *2020 data breach investigations report*. https://enterprise.verizon.com/resources/reports/2020-data-breach-investigations-report.pdf (Last accessed 05 May 2021).

Villasenor, J. D. (2011). *Ensuring Hardware Cybersecurity*. Center for Technology Innovation at Brookings.

Vlajic, N., & Zhou, D. (2018). IoT as a Land of Opportunity for DDoS Hackers. *Computer*, *51*(7), 26–34. https://doi.org/10.1109/MC.2018.3011046.

Whittaker, Z. (2013). *What Google does when a government requests your data*. https://www.zdnet.com/article/what-google-does-when-a-government-requests-your-data/ (Last accessed 05 May 2021).

Yang, F., & Xu, J. (2018). Privacy concerns in China's smart city campaign: The Deficit of China's Cybersecurity Law. *Asia & the Pacific Policy Studies*, *5*, 533–543. https://doi.org/10.1002/app5.246.

Yang, Y., Zheng, X., Guo, W., Liu, X., & Chang, V. (2019). Privacy-preserving smart IoT-based healthcare big data storage and self-adaptive access control system. *Information Sciences*, *479*, 567–592. https://doi.org/10.1016/j.ins.2018.02.005.

Yuan, E. (2020). *A message to our users*. https://blog.zoom.us/wordpress/2020/04/01/a-message-to-our-users/ (Last accessed 21 May 2021).

Zaveri, P. (2020). *Microsoft Teams now has 75 million daily active users, adding 31 million in just over a month*. https://www.businessinsider.com/microsoft-teams-hits-75-million-daily-active-users-2020-4 (Last accessed 21 May 2021).

CHAPTER 8

Role of artificial intelligence in fast-track drug discovery and vaccine development for COVID-19

Alka Bali[a] and Nishu Bali[b]
[a]University Institute of Pharmaceutical Sciences, UGC Center of Advanced Study, Panjab University, Chandigarh, India
[b]Department of Computer Sciences, Chitkara University, Patiala, India

1 Introduction

Recent times have seen a surge of reports employing computational methods and artificial intelligence (AI) for drug design. This stems from the fact that big data or data sets from modern biological techniques have become huge, complex and multidimensional and the classical computational methods do not have the capability to process them with acceptable levels of accuracy. Machine learning can accurately predict several compound properties like target binding affinity, membrane transport properties, toxicity aspects, pharmacokinetic or pharmacodynamic parameters without the need for an actual synthesis of the compounds or their evaluation.

Coronavirus Disease 2019, also known as COVID-19, was reported first from China in December 2019 and today, it has become a major pandemic currently afflicting 213 countries. The World Health Organization had pronounced this disease as a global public health emergency on January 30, 2020, presenting a huge risk for countries with vulnerable healthcare arrangements (Fig. 1) (Sohrabi, Alsafi, & O'Neill, 2020).

This pandemic has more than 22 million cases worldwide (6.48 million active cases), claiming more than 7.78 lakh lives in a short span of eight months, as per the data available on August 18, 2020 (Worldometer, 2020).

2 Artificial intelligence in COVID-19

Artificial intelligence is a powerful technique that has practically invaded all major spheres of our lives today. It makes a computer study, learn from the

Novel AI and Data Science Advancements for Sustainability in the Era of COVID-19
https://doi.org/10.1016/B978-0-323-90054-6.00006-4

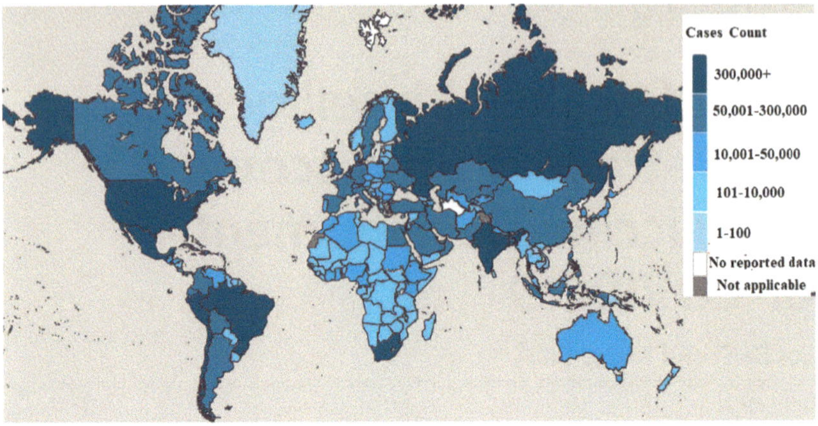

Fig. 1 Global data on COVID-19 cases available from the World Health Organization. As of August 18, 2020, 21,732,472 cases of COVID-19 have been reported worldwide, including 770,866 fatalities.

existing dataset and their patterns, "think" and apply the acquired knowledge to make certain predictions. Today AI has profoundly advanced the field of healthcare through the introduction of a greater degree of automation, convenience and higher accuracy and precision in diagnostic and therapeutic care (Davenport & Kalakota, 2019). With the introduction of robotic technologies, medical personnel are being relieved of usual routine tasks, thus allowing them to concentrate more on more crucial aspects of patient care, but the robots are actually assisting surgeons during surgical procedures. In some cases, computer algorithms are actually outmatching the capabilities of radiologists in the detection of malignancy in tumors. Automated disinfection of patient rooms, collection and transport of blood samples, patient sample storage, and analysis are a few areas where AI makes its presence felt by reducing the risks posed to patients and healthcare personnel via human intervention. Industry 4.0, often referred to as the 4th industrial revolution, represents the heralding of the age of digitization led by various types of disruptive technologies, encompassing greater automation through cyber-physical systems (enabling autonomous information exchange), Industrial Internet of Things (IoT), Internet of Medical Things (IoMT), intelligent networking of machines, information and communication technologies (ICT), Drone technology and smart machine-driven decision-making processes (Alcácerac & Cruz-Machado, 2019; Oztemel & Gursev, 2020).

As the world grapples with the enormous spread of this disease coupled with virtually nonexistent remedial measures insight, artificial intelligence-based tools are being employed for tackling the problem at various levels (Fig. 2) and such AI-based applications in the COVID-19 pandemic have been widely reviewed (Kumar, Gupta, & Srivastava, 2020; Lin & Hou, 2020; Naudé, 2020; Nguyen, Waurn, & Campus, 2020; Ting, Carin, Dzau, & Wong, 2020; Vaishya, Javaid, Khan, & Haleem, 2002; Vashisht et al., 2020).

Deep learning tools are being employed to aid in accurate diagnosis by helping analyze computational tomography (CT) scans, X-ray images predicting the advancement of the infection (Jin et al., 2020; Li, Zhang, Goncalves, et al., 2020; Narin, Kaya, & Pamuk, 2021; Wang, Kang, & Ma, 2020; Wang, Zha, Li, et al., 2020) and (or) mortality based on the patient's clinical data. For example, Dhiman et al. have reported a multiobjective optimization- and deep learning-based technique for detecting COVID-19 infection in patients employing chest X-ray images (radiographs) (Dhiman, Chang, Singh, & Shankar, 2021). COVID-19 is known to cause pulmonary parenchymal ground-glass opacification in the lungs

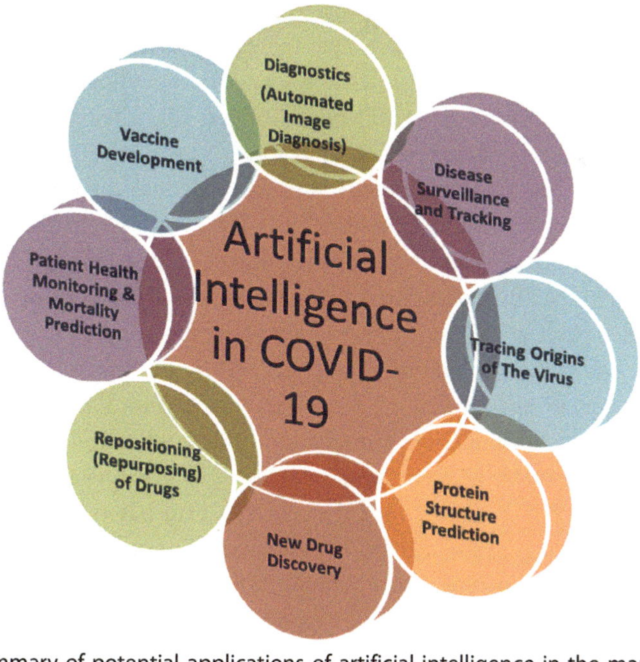

Fig. 2 Summary of potential applications of artificial intelligence in the management of COVID-19.

with subsequent consolidation on the advancement of the disease. In this study, eleven different types of CNN (convolutional neural network) based models have been developed, cross-validated and tuned using MOSHO (multiobjective spotted hyena optimizer). The reported models have generated good performance metrics, thus showing their usefulness in the real-time classification of COVID-19 disease. In another study, Abdel-Basset et al. have proposed a hybrid approach based on the thresholding technique in order to address the image segmentation problem for chest X-ray images (Abdel-Basset, Chang, & Mohamed, 2020). In this work, the integration of the slime mold algorithm (SMA) with the whale optimization algorithm has yielded results superior to several other algorithms.

Wang et al. have based their development of a fully automatic deep learning system for diagnosis and prognosis of COVID-19 on computed tomography images from a selected group of COVID-19 patients (Wang, Zha, et al., 2020). A semi-supervised few-shot segmentation (FSS) approach has been applied to carry out accurate and efficient segmentation of SARS-CoV2 infection (FSS-2019-nCov) with potential applications in limited data situations on a small dataset of annotated lung CT scans (Abdel-Basset, Chang, Hawash, Chakrabortty, & Ryan, 2021). Further, the deep learning system could not only distinguish COVID-19 from other types of pneumonias but was also able to segregate the patients into high- and low-risk propensity. AI-based methods are being employed for tracking COVID-19 spread over time and place (Chen, Lu, Chang, & Liu, 2020; Metabiota Epidemic Tracker, 2020; Rao & Vazquez, 2020; Tuli, Tuli, Tuli, & Gill, 2020). AI is further being used for disease surveillance, to scan public spaces for people with potential COVID-19 infection, and to enforce various social distancing measures or imposition of lockdown (Gurucharan, 2020; Maslan, 2020). Researchers have employed a mydriad of disruptive technologies such as AI, IoT, IoMT, virtual reality, big data, blockchain, 5G, robots, etc. to devise an intelligent framework for detection and prevention of the spread of COVID-19 disease (Abdel-Basset, Chang, & Nabeeh, 2021). The artificial intelligence-enabled mobile application called Aarogya Setu launched by the Government of India, is a good example of which helps the registered users to check their safety status based on whether they have crossed paths with Covid-19 positive patients. Thermal cameras possessing AI-based multisensory technology are being deployed in hospitals, airports, and public places to overcome the disadvantages of manually operated cameras. Artificial intelligence is also playing a very important role in tracing the origin of the novel coronavirus. Nguyen et al. have used artificial intelligence (AI) to carry out an investigation of the origin of the COVID-19 virus (Nguyen et al., 2020). The researchers have used raw genomic sequences of the virus, exceeding three hundred in number, collected from infected cases across different countries and analyzed by unsupervised clustering methods. Their results support the scientific hypotheses postulating bats and pangolins as the probable hosts for the SARS-CoV2 virus. Another similar study to track the virus origin has utilized 334 complete genomic sequences of the virus taken across the world, which have been lodged in the GenBank database. This analysis has been carried out based on unsupervised clustering methods, employing a hierarchic clustering algorithm twinned with DBSCAN (density-based spatial clustering of applications with noise), which has progressively narrowed down the search parameters, progressing from higher to lower levels of taxonomic classification. This study has concluded that SARS-CoV-2 belongs to the Riboviria realm or cluster (earlier represented by the coronavirus responsible for the Middle East respiratory syndrome or MERS) and progressive narrowing of cut-off parameters showed SARS-CoV-2 clustering with only two viruses. These included the bat CoV RaTG13 virus and the Guangdong pangolin CoV. Further pinning down of search suggested that SARS-CoV-2 virus groups exclusively with the bat. CoV-RaTG13 suggested that bats could be the most probable reservoir host of the novel coronavirus. The important role of AI in COVID-19 has been acknowledged by the WHO, creating a new technology access pool for the pandemic (The Pharma Letter, 2020).

3 Artificial intelligence in drug discovery

The most important application of AI in combating the SARS-CoV-2 virus is the development of potential treatment approaches in the arena of drug discovery and the development of vaccines. Considering that COVID-19 is a major health emergency, it is imperative that effective treatment measures be discovered in the shortest possible time frame. Recent times have seen a spurt in drug discovery institutes/ pharmaceutical industry collaborations with artificial intelligence platforms. For example, Scripps Research, with its drug repurposing collection ReFRAME has recently announced a collaboration with Repurpose.AI ActivPred artificial intelligence (AI) drug discovery platform to discover drug candidates effective against the novel coronavirus (The Science Advisory Board, 2020). The company, French AI startup Iktos, which is specifically engaged in applying artificial intelligence approaches for identifying novel drug candidates, has recently announced its collaboration agreement with a research institute in California, SRI International, to accelerate the development of novel antivirals against various types of viruses, including the influenza virus and the novel coronavirus (COVID-19).

4 Chemical structure input for data processing

Several repositories are available from which information relating to chemical structures of various viral proteins as well as drug-target interaction may be retrieved, e.g., the Protein Data Bank (PDB), NCBI PubMed, and NCBI PubMed Central (PMC), PharmGKB database, Therapeutic Target Database (TTD), etc. and these can be used for docking analyses or repurposing strategies. Several small-molecule libraries are there, from which structures of active molecules or approved/investigational drugs may be obtained, e.g., ChEMBL, DrugBank, drugCentral, GHDDI (virus-specific dataset), etc. The chemical structures of drugs with SMILES (Simplified Molecular Input Line Entry System) format can be extracted from DrugBank (Weininger, Weininger, & Weininger, 1989). Multistage text processing and data analysis of abstracts and full papers included in public repositories have been used to identify common host-protein targets between HIV-1 and COVID-19 infections by feeding queries built by employing Python scripts. Full-text articles or abstracts can be retrieved in machine-readable XML format from several of these repositories. Tarasova, Ivanov, Filimonov, and Poroikov (2002) have employed the Lingpipe-4.1.2 algorithm fine-tuned on GENIA corpus (Kim, Ohta, Tateisi, & Tsujii, 2003) to extract out protein sets and to identify overlapping proteins. They have further saved species-specific proteins related to the viruses under study and the host proteins out of this complete dataset by prior checking on UniProt API. Hu, Jiang, & Yin, 2020 have trained their multitask deep model consisting of shared layers and task-specific layers by a virus-specific dataset to carry out drug prediction against the COVID-19 virus. The binding sites have been predicted in the selected viral proteins by exploring critical input sequences by applying a nonparametric method occlusion. In another study, standard protein interaction networks have been utilized for identifying repurposing opportunities based on interactions of SARS-CoV-2 proteins with other proteins and the interaction networks between molecules in the cell (Kumar, 2020).

Generative deep learning approaches have been utilized for arriving at novel drug-like candidates (Zhavoronkov, Aladinskiy, Zhebrak, et al., 2020).

Active ligand complexed within the crystal structure of SARS-CoV main protease was extracted for carrying out ligand-based generation. A pocket algorithm has been used to annotate the binding site. Employing an X-ray structure of 4MDS from SARS-CoV M^{pro} (co-crystallized with a noncovalently bound ligand) as a template, a homology model has been constructed for SARS-CoV-2 3C-like protease-ligand complex using SWISS-MODEL. Further, a dataset of protease-binding molecules has been constructed by extraction from repositories and filtering out highly nondrug-like molecules (such as radicals, metals, hydrazines, etc.) via the application of Mild medicinal chemistry filters (MCFs). Several ML models, including generative autoencoders, generative adversarial networks, genetic algorithms, etc. were used to create novel molecular structures followed by optimization through reinforcement learning (RL) employing the reward functions based on scores for medicinal chemistry & drug-likeliness, active chemistry (utilizing self-organizing maps previously fine-tuned on protease peptidomimetics dataset), structure-fitting to receptor pockets, novelty (e.g., the penalty for similarity to existing molecules), etc. The low reward was given to molecules with structural alerts (Guex, Peitsch, & Schwede, 2009). In another study, generative and predictive models for de novo design of novel molecules against the SARS-CoV-2 3CL protease have been reported (Bung, Krishnan, Bulusu, & De Roy, 2020), based on a deep neural network. For preprocessing of data, Python (RDKit library) was used, and molecules were input in SMILES notation, followed by introducing chemical filters and removing SMILES strings longer than 100 symbols. The molecules in SMILE format were then used to train the generative model, which was then retrained with another dataset of protease inhibitors (ChEMBL) by transfer learning. Drug-target interaction information in various studies is generally used to build the drug-target network and this is generally based on binding affinity criteria like K_i, K_d, IC_{50}, or EC_{50} (should be low; in range of nM or at submicromolar concentrations). For example, Zhou, Hou, and Shen (2020) have employed the input criterion that these affinity values should not exceed 10 μM. In another deep learning approach for virtual drug screening (Zhang, Saravanan, & Yang, 2020a), the workers retrieved the viral RNA sequences from Global Initiative on Sharing All Influenza Data (GISAID) database (Shu & McCauley, 2017), which were then translated to their respective amino acid sequences employing a web tool (Translate).

5 Artificial intelligence in repositioning approaches

Considering the large-scale and rapid spread of COVID-19, it is increasingly recognized that there is a need for immediate discovery of drug treatment before any vaccine is developed. Hence, the major focus targeted towards mitigating the virus has been on the repositioning approaches, i.e., identifying new clinical uses for existing drugs approved by the US Food and Drug Administration (US-FDA) (Fig. 3). Drug repurposing is recognized as an efficient strategy for drug discovery which can significantly reduce the time and costs in comparison to the de novo drug discovery approaches and randomized clinical trials (Cheng, Hong, Yang, & Wei, 2017; Cheng, Murray, & Rubin, 2016). Simple and computational drug repurposing has been previously utilized to identify plausible drug candidates with the potential to be effective against other viral infections such as dengue, Ebola, ZIKA and influenza (Johansen, Brannan, Delos, et al., 2013; Madrid, Chopra, Manger, et al., 2013; Mani, Wadhwani, & Krishnamurthy, 2019).

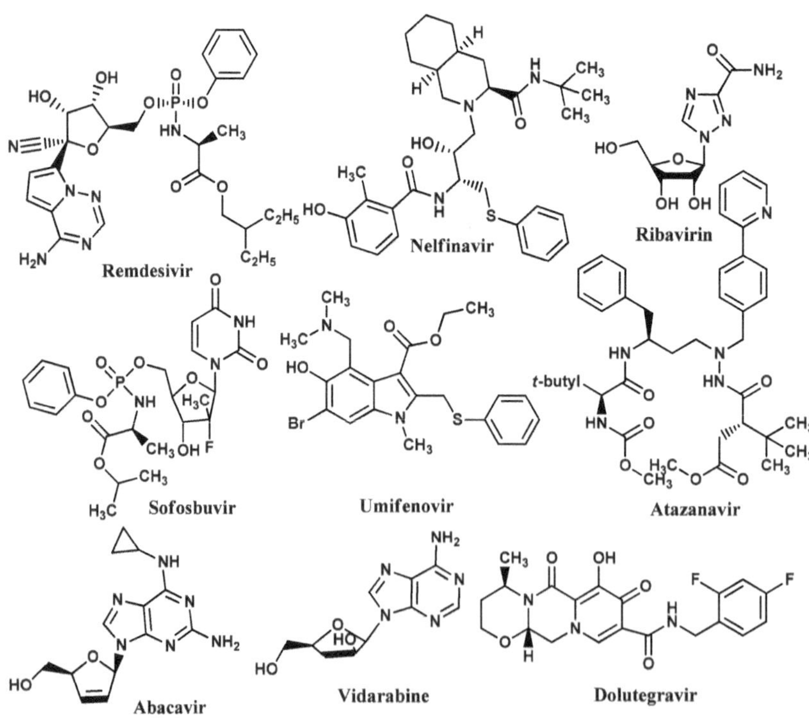

Fig. 3 Antiviral drugs repurposed against SARS-CoV-2 infection through artificial intelligence approaches.

Experimental repurposing approaches are relatively time-demanding and costlier compared to the computational approaches (Santos et al., 2020) and hence, it may be followed after zeroing on viable drug candidates through the application of computational approaches. Computational repurposing has earlier shown promising results against infections by the SARS-CoV and MERS-CoV viruses (Dyall, Coleman, & Hart, 2014; Nukoolkarn, Lee, & Malaisree, 2008) and following the SARS-CoV2 outbreak, computational repurposing is being applied in search of novel treatment strategies. One of the more widely used drugs in COVID-19 treatment has been hydroxychloroquine, an antimalarial drug, though its efficacy and safety aspects have been debated. This drug was identified through AI-based and big data platform (Medicircle, 2020) through analysis of drug pathways and interactions to identify targets for human/virus proteins. Further, this AI-based platform has carried out the evaluation of the clinical potential of drugs previously approved for the treatment of other diseases for postulating potential drug combinations for COVID-19. Pertinent suggestions include a combination of Chloroquine with an antiarthritic drug Tocilizumab to counter cytokine release syndrome, a well-known complication of COVID-19 (Zhang, Wu, Li, Zhao, & Wang, 2020). Other potential drug combinations that have been suggested include, a combination of chloroquine with the antiviral drug remdesivir, and a combination of hydroxychloroquine with a semisynthetic macrolide antibiotic and antiretroviral drug clarithromycin (Medicircle, 2020).

The phylogenetic analysis of fifteen human coronaviruses (HCoV) genome sequences has revealed that the novel coronavirus (SARS-CoV2) collected from patients at the beginning of the outbreak has a maximum (79.7%) sequence identity with respect to the SARS-CoV (Jaimes, André, Chappie, Millet, & Whittake, 2020; Zhou, Yang, & Wang, 2020). Further, two evolutionarily conserved regions of SARS-CoV-2, i.e., the envelope and nucleocapsid proteins, possess much higher sequence identities with respect to SARS-CoV (96% and 89.6%, respectively) (Jaimes et al., 2020). Thus, drugs that have previously shown promise against SARS-CoV should have the potential to be effective against this novel coronavirus as well.

Recently, network-based repurposing of drugs employing network proximity analyses of the drug targets and interactions between Human CoV and the host has been employed to prioritize sixteen potential anti-HCoV drugs that could be repurposed COVID-19 (e.g., melatonin, mercaptopurine, and sirolimus) (Zhou, Hou, & Shen, 2020). The prioritized drug molecules have been validated through enrichment analyses carried out with drug-gene signatures and the transcriptomics data in human cell lines induced by the human CoV. Other useful information has been obtained from the "Complementary Exposure" pattern, i.e., when targets of two drugs hit the sub-network between Human CoV and the host. However, they target separate neighborhoods in the network of the human interactome. This information has been utilized to identify three potential drug combinations for effectiveness in SARS-CoV2 infection, i.e., sirolimus and dactinomycin, mercaptopurine and melatonin, toremifene and emodin.

Advanced molecular simulations are being combined with AI-guided search for potentially active drug molecules to facilitate the identification of molecules that can be repositioned as prospective antiviral drugs (Koppal, 2020). Research by an AI drug discovery company BenevolentAI has utilized AI software to discover an approved rheumatoid arthritis drug, Baricitinib, having the capability to diminish the infectivity of the virus (Richardson, Corbellino, & Stebbing, 2020; Richardson, Griffin, & Tucker, 2020).

Recently, four small molecular drugs have been identified on the basis of their high binding affinity for the SARS-CoV main protease via high-throughput screening of eight thousand approved/experimental drugs and small molecules acquired from Drugbank. These include the fluoroquinolone antibiotic Prulifloxacin, anti-HIV-1 drugs Bictegravir and Nelfinavir, and the antiviral drug Tegobuvi (Li, Zhang, Wang, et al., 2020). Molecular similarity search was performed on the basis of the similarity in sequences of the structure-divulged molecules. Out of the resulting 690 compounds, only 50 were subjected to docking studies and the remaining compounds were discarded based on their toxic profiles, strong side effects, antitumor properties, etc. All these drugs have been widely employed in clinical practice and have been proposed as promising candidates for the treatment of the infection from 2019-novel coronavirus. Nelfinavir has also been predicted to be a prospective inhibitor of SARS-CoV 3C-linked main protease ($3CL^{pro}$ or M^{pro}) by another integrative computational-based approach based on a combination of homology modeling, docking and binding free energy calculations (Xu, Peng, Shi, et al., 2020). In this report, the SARS-CoV main protease structures have been modeled, taking as a template the SARS homolog (PDB ID: 2GTB). Fifteen drugs were initially selected from docking score and three-dimensional (3D) similarity analyses and these were again docked with ten additional models of the protease enzyme. Six drugs have been chosen for further studies. These include Praziquantel,

Pitavastatin, Zopiclone, Nelfinavir, Eszopiclone and Perampanel. Binding free energy calculations on four of these drugs employing MM/GBSA and SIE approaches retrieved nelfinavir as the most promising candidate (Xu et al., 2020).

As mentioned previously, two inhibitors of HIV-1 protease enzyme, i.e., lopinavir and ritonavir had been identified as inhibitors of the SARS-CoV M^{pro} and these have also been found to be clinically effective against SARS-CoV2 (having 96% similarity to SARS-CoV) (Nukoolkarn et al., 2008). Further, the three-dimensional structure of the binding pocket where lopinavir/ritonavir interact with the enzyme is also conserved (the conserved amino acids being Threonine (Metabiota Epidemic Tracker, 2020)-Asparagine (Abdel-Basset, Chang, & Nabeeh, 2021) and Asparagine; numbering as per SARS-CoV protease) between the main protease of both SARS-CoV and SARS-CoV2. Amino acids Threonine (Metabiota Epidemic Tracker, 2020), Threonine (Gurucharan, 2020), and Asparagine[119] have been predicted to bind to the two drugs through hydrogen bonding. Based on these results, ten clinical drugs have been identified from the DrugBank database that can forge hydrogen bonds with conserved amino acid residues in the binding pocket of the enzyme SARS-CoV2 M^{pro} along with the possibility of higher tolerance to resistance mutations (Liu & Wang, 2020). There is a significant amount of data on molecular mechanisms of HIV-1 infections and several anti-HIV drugs have been found to inhibit the SARS-CoV-2 virus. Considering this relationship, text and data mining workflow have been used to identify as many as 46 targets. The study has shown that the two viruses share some common molecular pathways for the generation of their inflammatory, immune and cell cycle responses (Tarasova et al., 2002). Studies have evoked that the virus can enter the human cells by binding with ACE2 (angiotensin-converting enzyme 2), a cell-surface molecule (Lu & Sun, 2020; Zhao et al., 2020) and this is crucial for the human cell entry of the virus. The study by the BenevolentAI group had adapted its search to the available data on ACE2. In this regard, the enzyme AAK1 (adaptor-associated protein kinase 1), which is responsible for the regulation of endocytosis (entry of material into the cells) was identified as a possible disease target. Baricitinib emerged as the final drug out of 378 known AAK1 inhibitors included in the study, and the researchers have proposed a clinical investigation of the drug against the virus (Stebbing, Phelan, & Grifn, 2020). Accelerated clinical trials are proposed to be conducted on baricitinib to test its effectiveness in COVID-19 by the National Institute of Health, UK. This drug is presently being explored by several other countries including Canada, Italy, and the US (Smartcity, 2020) and clinical studies also demonstrate its usefulness along with other repurposed drugs for COVID-19 (Sodani, Mucci, & Girolimetti, 2020).

Recently, another computational AI model has been reported for drug repurposing for mitigation of SARS-CoV-2 infection. For screening a database of more than a hundred approved antiviral drugs, the AI model has followed the approach of drug-target interaction prediction, i.e., it retrieved drugs effective against viruses having a genomic similarity with respect to the novel coronavirus. The model initially computes structural complementarity of the chemical structures of the drugs and the genomic structures of different viruses along with SARS-CoV-2 and then considers the diachronic data relating to the efficacy of the drugs against various types of viruses. All the drugs selected by the developed model (based on structural complementarity to viruses having genomic similarity to SARS-CoV-2) as prospective cures for COVID-19, i.e., sofosbuvir, remdesivir, umifenovir and ribavirin are presently undergoing clinical investigation against COVID-19 infection. Furthermore, AI could also offer a solution to the problem of mutation of the novel coronavirus as drugs effective against the original strain that surfaced in December 2019. These are considerably different from those effective against the more recent strains encountered from June 2020 onwards (The Indian Practitioner, 2020).

Deep learning-based models have been employed to carry out prediction of binding affinities based on SMILES chemical sequences of the target protein along with the sequences of amino acids (FASTA). Drugs like Atazanavir, Remdesivir, Kaletra (Lopinavir/ritonavir combination), Rapamycin and tiotropium bromide have been identified as potential inhibitors of the SARS-CoV2 virus. Amongst these drugs, remdesivir has recently been approved by US FDA for use in COVID-19 patients (Narayana et al., 2020). Beck, Shin, Choi, Park, and Kang (2020) have also predicted the antiretroviral drug atazanavir as the best candidate for inhibition of the SARS-CoV2 3C-like protease enzyme followed by efavirenz, ritonavir and dolutegravir.

Nguyen, Gao, and Chen (2020) have applied a Deep Learning approach based on mathematical techniques (MathDL) for the identification of drugs with inhibition potential against 3CL like protease enzyme. For this purpose, they have employed a homology model of 3CL like protease, prepared to take the crystal structure of SARS-CoV 3CL protease as the template. Training sets were prepared from eighty four inhibitors of SARS coronavirus protease enzyme taken from the ChEMBL database. Another general dataset containing more than fifteen thousand ligand-protein complexes was taken from the PDBbind database. Based on these, deep learning models have been developed by combining algebraic topology with (i) deep convolutional neural networks (CNNs) and (ii) deep multitask CNNs. Evaluation of a set of 1465 FDA-approved drugs for their binding affinities has identified a list of fifteen prospective drug candidates included in the DrugBank dataset, the top-ranking candidate being Bortezomib followed by flurazepam and ponatinib in this order (Nguyen, Gao, & Chen, 2020).

In another similar study (Zhang, Saravanan, & Yang, 2020c), the RNA sequences of 2019-nCoV from 18 patients lodged in a public domain database (GISAID) were first retrieved, followed by a translation into corresponding protein sequences and multiple sequence alignment. This study indicated 3C-like protease as a key therapeutic target. Further, an indigenously built homology model of the 3CL protease was taken and a deep learning model was employed for performing large-scale virtual screening to identify protein-ligand interacting pairs. Drug screening has been carried out from four chemical compound databases using a deep learning model founded on the Dense Fully Convolutional Neural Network (DFCNN). The potential ligands identified from this study include the small molecule drugs like Meglumine (amino sugar), D-Sorbitol (sugar alcohol), Adenosine nucleoside, D-Mannitol (sugar), Sodium gluconate (chelating agent), Ganciclovir (antiviral drug) & Chlorobutanol (sedative) and peptide drugs, (comprised of lysine, isoleucine, and proline).

PolypharmDB is a recently developed repository of clinically evaluated molecules along with their multiple pharmacological profiles for identifying their repurposing opportunities. Deep learning engine, MatchMaker™, which has been trained on millions of reported drug-target interactions in humans, has been used along with PolypharmDB MatchMaker to perform a rapid investigation of human target proteins and various viral proteins for potential therapeutic significance in COVID-19 (Kurji, 2020). This study has resulted in a set of molecules with potential of interacting with COVID-19 at various targets.

Hu, Jiang, & Yin, 2020 have also reported a prediction study. They first fine-tuned their pre-trained multitask deep model using a dataset of marketed drugs that have demonstrated in vitro inhibition of the novel coronavirus through in vitro studies. This re-trained model was then used to predict the affinities of a dataset of 4895 commercially available drugs against eight prospective protein targets present in SARS-CoV-2. Ten promising drug candidates have emerged from this study. Amongst these, abacavir, a powerful reverse transcriptase inhibitor (anti-HIV), was predicted to possess a high affinity of binding towards several proteins of SARS-CoV-2, including RdRp, 3CL main protease, papain-like protease and helicase. Another anti-HIV drug, darunavir, a protease inhibitor, has also been predicted to target 3CL protease, RdRp and papain-like protease enzymes with nanomolar affinity. Other drugs include the antiasthmatic drug Fiboflapon sodium (predicted to have a potential affinity to Papain-like protease), daclatasvir used against Hepatitis-C virus predicted to inhibit RdRp, almitrine (respiratory stimulant) with a predicted affinity towards 3C-like proteinase and RdRp, Itraconazole with a predicted affinity towards 3C-like proteinase, roflumilast (antiinflammatory) with a predicted affinity towards Papain-like protease.

RdRp and metoprolol (antihypertensive) (Fig. 4) with a predicted affinity towards Papain-like protease (Hu, Jiang, & Yin, 2020).

A combination of network algorithms and human curation has been used to search integrated knowledge graphs and the information has been used for identifying drug repurposing opportunities against SARS-CoV-2. Eight potential repurposing opportunities have been identified in the form of drug classes (e.g., cathepsin L inhibitors; calmodulin inhibitors) and individual drugs like fostamatinib. In this study, a semantic knowledge graph was constructed in Neo4j, including the data of already existing, approved drugs (Skelton, Alsobhe, & Anastasi, 2020). Various data sources included for this purpose include UniProt (protein database), DrugBank (drugs), Monarch Disease Ontology (MONDO) (diseases/disorders), Drug Central (drugs), OMIM (human genes and genetic phenotypes), DisGeNET (database related to gene-disease associations). A literature search from the selected databases was used to identify existing know-how on genetic and proteomic mechanisms of pathogenesis in *Homo sapiens* and SARS-CoV-2 and this knowledge were used to devise semantic queries.

A high-throughput docking approach has been applied (Cavasotto & Filippo, 2020) employing a novel quantum mechanical grading for *silico* repositioning of drugs for COVID-19. In this study, they have screened a chemical library of more than eleven thousand compounds comprising of drugs approved by the US-FDA and molecules under clinical evaluation against three target proteins of the SARS-CoV-2. These proteins include the spike protein called the S-protein, the 3C-like proteinase, and the Papain-like protease. The docking library of 11,552 compounds has been constructed by coalescing molecular subsets obtained from 4 chemical compound libraries, i.e., (i) ChEMBL, which included molecules reaching at

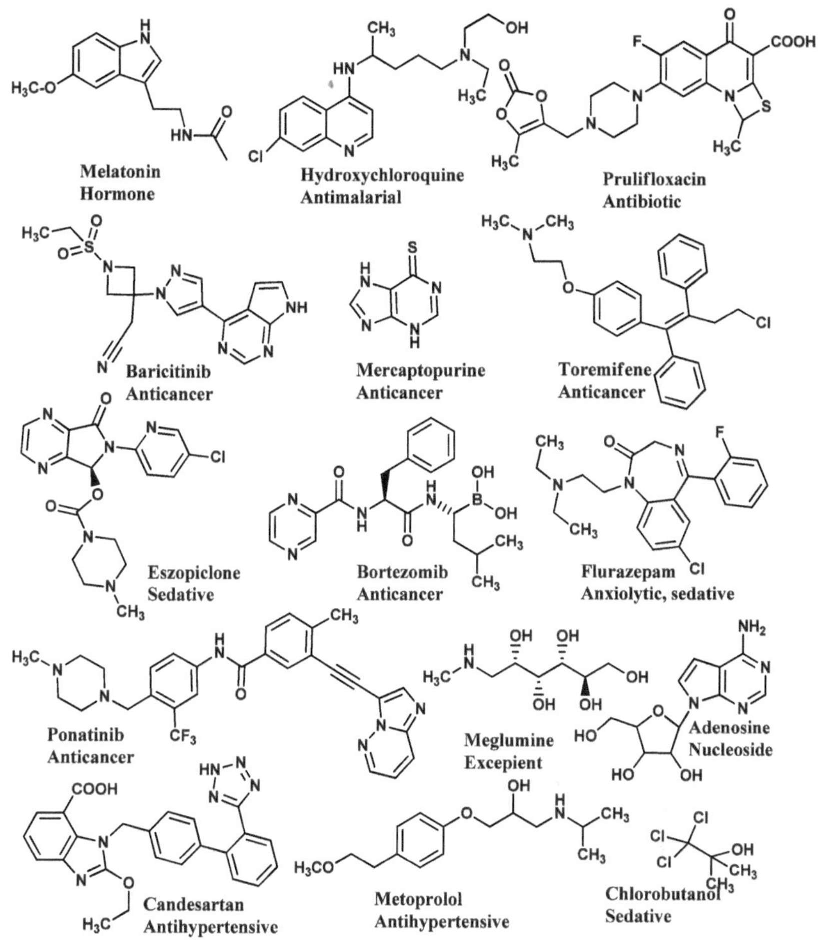

Fig. 4 Drugs not previously documented as antivirals repurposed against SARS-CoV-2 infection through artificial intelligence approaches.

least one of the phases of clinical trials; (ii) DrugBank, which included all approved as well as investigational drugs; (iii) DrugCentral database comprising of approved drugs and (iv) the library of FDA approved drugs from Selleck Chem. Potential inhibitors identified by this study include the antiviral drugs sovaprevir, samatasvir, elbasvir and saquinavir and several other drugs with diverse indications like the AT_1 receptor antagonist candesartan, antidiuretic drug desmopressin, thrombin inhibitor flovagatran and the antineoplastic drug brilacidin (Cavasotto & Filippo, 2020).

Another AI platform has been established to repurpose old drugs against COVID-19 by using two different learning databases; first, comprising compounds reported to have activity against SARS coronaviruses (SARS-CoV and SARS-CoV-2), HIV, influenza virus, and second comprising the known inhibitors of 3C-like main protease (Ke, Peng, Yeh, et al., 2020). activity against a feline strain of coronavirus and the results were used for relearning of the AI model. Eighty marketed drugs were predicted to have potential against the virus and amongst these, eight drugs, i.e., bedaquiline, celecoxib, gemcitabine, clofazimine, tolcapone, brequinar, conivaptan, and vismodegib, actually demonstrated in vitro activities against the virus responsible for feline infectious peritonitis (FIP). Further, boceprevir, chloroquine, homoharringtonine, tilorone, and salinomycin have also been predicted to be active through the developed AI model (Ke et al., 2020).

6 Artificial intelligence for accelerating computer modeling

Besides the repositioning approaches, which appear to be the most promising, efforts are also afoot to screen a large volume of small molecules for their potential to interact with the virus, particularly the viral proteins, to disrupt its activity. Computer-aided molecular design has been the most common approach in recent times for designing/screening molecules against macromolecular targets involved in diverse types of diseases. However, for an unknown virus-like SARS-CoV-2, where no particular lead molecule is known yet, one needs to screen billions of small molecules against more than twenty known proteins in the virus to assess their potential effectiveness in COVID-19 infection. The utilization of high-performance computational tools like supercomputers coupled with AI and machine-learning can considerably speed up this process by enabling the prediction of key viral proteins and quick screening of billions of molecules to arrive at pertinent drug candidates (Zhavoronkov, 2018). Four categories of structural proteins present in the SARS-CoV-2 virus are important, i.e., the spike proteins (S), nucleocapsid proteins (N), envelope proteins (E) and membrane proteins (M) (Lu, Zhao, Li, et al., 2020; Zhavoronkov et al., 2020). Further, some nonstructural proteins (NSPs) are there which are crucially involved in the viral pathogenesis. These include the C30 endopeptidase, more commonly referred to as 3-chymotrypsin-like protease (called 3C-like protease or 3CLPro or MPro (the main protease), or nsp5) & the papain-like protease (PLpro, component of nsp3). At the human level, research is being directed towards the angiotensin-converting enzyme 2 (ACE2), a receptor known to facilitate the entry of the virus into the cells of the host. Several studies suggest that the SARS-CoV-2 utilizes this ACE2 receptor for gaining entry into the host cells and the Transmembrane protease serine 2 gene (TMPRSS2) is used for S-protein priming (Hoffmann, Kleine-Weber, Schroeder, et al., 2020).

Recently, a collaborative work has been conducted by integrating artificial intelligence and machine-learning tools, i.e., Deep-Learning-Guided Adaptive Molecular Simulations for Protein Folding (DeepDriveMD) with drug docking based on physics concepts and molecular dynamics simulation tools to fast-track the discovery of promising molecules against the virus (Dubrow, 2020). Millions simulations have been run as a part of this hybrid AI- and physics-based modeling-based project undertaken with some of the most powerful supercomputers like Frontera, Longhorn, Summit, Theta and Comet at multiple laboratories, which can simulate lakhs of molecules per hour. The results from these simulations have been used to train the machine learning system to identify crucial molecular parameters for activity. This can be utilized to screen other molecules for their potential activity. This work has yielded 30 molecules (out of a billion molecules taken in the initial stage of screening) with the potential to block a crucial enzyme involved in viral replication, adenosine diphosphate ribose 1″ phosphatase (ADRP) (Dubrow, 2020).

Besides the repurposing approaches or screening of existing chemical databases for potential hits against salient viral proteins, generative chemistry approaches or de novo approaches are also being explored to encompass a virtually limitless number of chemical structures. Such deep generative models employ huge datasets for fine-tuning to carry out in silico designing of de novo chemical structures with required pharmacological or biochemical characteristics (Zhavoronkov et al., 2020). Recently, an HPC wire report has suggested the potential of quantum machine learning, a combination of machine learning and quantum information processing approaches for quicker and economical generation of novel complex compounds for COVID-19 (HPC Wire Staff Report, 2020). Huge investments are being made in this area by research institutes and researchers are employing machine learning methods to gain additional knowledge on "molecular aspects" of the virus and the mechanism behind the harmful effects of the virus on lungs (Rosenblatt, 2020). The study on the viral effect on lungs has been directed towards identifying a protein network interacting with the COVID-19 virus and enabling it to spread further to determine which proteins could possibly be targeted to arrest the viral activity through drug treatments.

Currently, work is in progress to build an all-encompassing, all-atom model of the envelope of SARS-CoV-2 coronavirus, approximated to be containing more than two hundred million atoms using the Frontera supercomputer of Texas Advanced Computing Centre (The Science Advisory Board staff, 2020). This work is on the lines of a previous work in which an all-atom simulation of the influenza virus envelope had been reported (Durrant et al., 2020). This information could be useful for the scientists working on the design of new drugs against SARS-CoV-2 and can also shed some light on the mechanism of interaction of current drugs and new potential drug combinations with the virus.

7 Artificial intelligence: De novo design of novel small molecules

Recent advancements in artificial intelligence (AI) & machine learning enable efficient mining of existing knowledge to explore an unlimited volume of chemical space. This can be used to develop novel small molecular agents or new chemical entities (NCEs) having required pharmacological and physicochemical properties (Popova, Isayev, & Tropsha, 2018; Stokes et al., 2020). This approach has been applied recently to design novel small molecules to target SARS-CoV-2. One research group has employed generative and predictive models based on a deep neural network for carrying out de novo designing of novel chemical molecules for inhibiting the 3CL protease, an enzyme (biocatalyst) required for the growth of the virus (Bung et al., 2020). This generative model trained on a dataset consisting of around sixteen lakh molecules possessing drug-like properties (ChEMBL) was re-trained using another dataset consisting of 2515 molecules with inhibitory properties against protease enzymes by transfer learning (Gaulton, Bellis, Bento, et al., 2012). Further, the generative model was modulated by reinforcement learning to generate molecules having the desired properties. These small molecules generated through artificial intelligence were filtered on the basis of various physicochemical parameters and docked within the 3CL protease model (PDB ID: 6LU7). This study has generated thirty-one potentially active candidate compounds to be synthesized and evaluated against SARS-CoV-2. Two of these small molecules have also shown high similarity with respect to Aurantiamide, an antiviral compound of herbal origin, obtained from *Baphicacanthus cusia*, an extensively used herb in the Chinese system of medicine for the management of colds, flu and hyperthermia (Zhou, Yang, Feng, et al., 2017).

Another approach based on an advanced deep Q-learning network coupled with the fragment-based drug designing (ADQN-FBDD) has been employed in amalgamation with variational autoencoder (KL annealing & circular annealing) with the aim to generate lead molecules targeting the chymotrypsin-like protease (3CL main protease of SARS-CoV-2) (Verma & Bansal, 2020). This method employed SBOP (a structure-based optimization policy) in reinforcement learning. This work has generated three molecular structures which can potentially inhibit viral growth. Amongst these, one of the molecules bears functional similarity to the drugs remdesivir, umifenovir and elbasvir, which the US FDA approves for use in COVID-19 patients during emergency situations.

Insilico Medicine has also reported utilizing its previously validated generative chemistry approaches to design novel drug-like inhibitors of COVID-19 with SARS-CoV-2 main protease (Mpro) as the target protein. These include crystal-derived pocked-based generators, generation based on homology modeling, and ligand-based approaches (Zhavoronkov et al., 2020).

Generation of potential lead compounds against SARS-CoV-2 3CL (PDB ID: 6LU7) has also been carried out by utilizing the ADQN-FBDD approach (Tang et al., 2020). This was followed by a structure-based optimization policy (SBOP) to obtain forty-seven top lead compounds, which have been assessed through molecular docking simulations.

Two novel molecule candidates have been generated by deep generative models by Innoplexus that can stop the growth of the virus by targeting a specific virus protein. Preliminary results from in silico predictions of ADME profile and docking studies have demonstrated an increase in efficacy over existing drugs and further testing is being carried out (Medicircle, 2020).

8 Artificial intelligence in protein structure prediction

Information about viral protein structures is crucial to the understanding of protein functions and mechanisms of their inhibition by drugs or vaccines, but the experimental determination of protein structures employing techniques like X-ray crystallography, solid-state NMR, etc. is a time-consuming process. Moreover, these techniques will not be applicable if the pure isolated form of protein is unavailable for analysis. Hence, computational approaches are employed to predict protein structures either from sequences of component amino acids or through "template modeling" based on already existing structures of similar proteins.

Deep learning approaches of artificial intelligence have been utilized to carry out structure-based predictions of several proteins associated with SARS-CoV-2. AlphaFold system by Deep Mind (a division of Google Inc.) (Senior, Evans, & Jumper, 2020), is based on ResNet architecture (Yu & Koltun, 2016) and focuses on a template-free protein modeling approach, i.e., it can predict structure for those proteins which do not have any known related structure. It uses the sequences of component amino acids and MSA (alignment of multiple sequences) based on the features extracted from similar amino acid sequences to predict the spatial distance and angular alignment amongst the amino acid residues. This system has released structure predictions of six under-studied proteins associated with SARS-CoV-2, including the SARS-CoV-2 membrane protein, Nsp2, Nsp4, Nsp6, protein 3a, and the C-terminal domain of Papain-like protease (Deep Mind, 2020). Their system had also provided accurate structure prediction of the spike protein of the SARS-CoV-2, the experimentally determined structure of which is lodged in PDB (Protein Data Bank) (RCSB Id: 7BZ5).

The transform-restrained Rosetta (trRosetta) pipeline system also uses a dilated ResNet architecture for the prediction of the structures of the proteins mentioned above along with the ORF6, ORF7b, ORF8, and ORF10 proteins.

9 Artificial intelligence in vaccine development

Besides the very important role of artificial intelligence in designing drugs against COVID-19, attempts are now being directed towards the utilization of AI programs to facilitate the development of the COVID-19 vaccine. Virus-neutralizing antibodies are produced by the body via B-cells (humoral immunity) and T-cells (cellular immunity). A subtype of T-cells, termed memory cells are responsible for the recognition of an antigen of the eliminated pathogen, and any type of re-exposure to the pathogen quickly activates additional effector T-cells. All these processes are utilized for designing vaccines. The major histocompatibility complex proteins (MHC I and MHC II proteins) are the helper proteins that present the binding regions of the antigens, termed as epitopes, to the antibodies, B- or T-cells for binding and their consequent neutralization. AI programs allow the identification of epitopes, i.e., the viral antigenic portions that have a greater probability of exposure on the infected cell surface and can attach to the antibodies. One example is that of an interactive AI platform, "Epitopes world" intended to make predictions for possible vaccine targets and thus reduce the time taken and the expense involved in creating vaccine candidates (Epitopes World, 2020). This platform has been designed on an algorithm called "CAMAP" that had been originally applied for cancer immunotherapy.

Magar et al. have devised a machine learning model to explore antibodies to neutralize the antigens (Magar, Yadav, & Farimani, 2021). A training dataset (VirusNet) was prepared consisting of more than nineteen hundred previously reported sequences of antigen-antibodies from several allied diseases, including HIV, influenza, SARS and Ebola. Graph featurization has been used with various machine learning methods for the screening of a large number of hypothetical antibody sequences ranging in thousands. Eight antibodies have been proposed with potential against COVID-19. The stability of the candidate antibodies has been finally assessed through a combination of bioinformatics, structural biology, and Molecular Dynamics (MD) simulations.

Fast et al. have utilized previously-developed neural networks for the prediction of major histocompatibility complex protein presentation, i.e., NetMHCPan4 and MARIA for identification of more than four hundred potential epitopes on the T-cells that can be presented by the major histocompatibility complex proteins (MHC I & MHC II), and two epitopes on the S-protein, targeting B-cells (Fast, Altman, & Chen, 2020). Further, 68 genomic variants of the SARS-CoV-2 have been examined to investigate the mutational characteristics of the virus in order to recognize which viral parts are less susceptible to phylogenesis. It has been observed that mutations in the case of the virus are more probable in those regions which have a strong antigenic presentation and the S-protein epitopes should make good vaccine targets as no nearby mutations have been found near the specific domain of the spike protein involved in receptor binding (Fast et al., 2020).

Reverse vaccinology has revolutionized the vaccine design approach in recent times. It focuses on identifying promising candidates for vaccine development through bioinformatics-based analysis of the genome of the pathogen. A recent study has employed reverse vaccinology and machine learning (ML) approaches for designing of vaccines against COVID-19 (Ong, Wong, & Huffman, 2020). In this study, Vaxin RV (a web-based program) has been employed along with Vaxign-ML (Ong, Wang, & Wong, 2020), a machine learning tool developed by the researchers to enhance prediction accuracy, for predicting the viral proteins which can serve as effective targets for COVID-19 vaccine. In this study also, the spike protein (or S-protein) has been found to demonstrate the highest protective antigenicity score, thus, emerging as the best vaccine candidate. Additionally, five nonstructural proteins or NSPs, i.e., nsp3, 3CL-pro, and nsp8–10, have been identified as vaccine targets. The spike protein and the nonstructural proteins-3 and 8 (nsp3 & nsp8) have been predicted by Vaxign-ML to be the antigens with high protective antigenicity. The nsp3 protein has been more conserved amongst SARS-CoV-2, SARS-CoV, and MERS-CoV than amongst fifteen coronaviruses that can infect humans/animals; therefore, nsp3 protein has been selected as the second-best or alternative vaccine candidate to the S-protein.

Malone et al. have reported AI-enabled prediction of blueprints for designing universal vaccines for SARS-CoV-2 (Malone, Simovski, Moliné, et al., 2020). These are designed to include a reasonably extensive repertory of T-cell epitopes to ensure that the global population is broadly covered and protected. In this study, the presentation of the antigen to the host cell surface infected by the virus and predictions regarding immune response generation from *NEC Immune Profiler* has been used to profile the complete proteome of SARS-CoV-2 followed by the use of robustness Monte Carlo as well as digital twin simulations to identify a subset of epitope hotspots that have the maximum probability of being immunogenic amongst a broad spectrum of HLA allele types. These could be used for the formulation of the SARS-CoV-2 vaccine that can provide wide coverage of the population around the globe.

10 Summary

The novel coronavirus disease (COVID-19) has spread to more than 200 countries worldwide, thus assuming the proportions of an enormous global pandemic. Virtually, all spheres of life have been affected by the disease and hence, it is attracting the attention of researchers not only in pharmaceutical academia/industry/health profession but also in AI technology. AI has been utilized extensively for numerous extra-clinical applications in modeling viral transmission/spread of the disease, monitoring quarantine measures, and general data analysis. The health/pharmaceutical professionals are the frontline runners in this race as they are trying to contain this disease through novel drug discovery, repositioning of existing drugs and vaccine development. Artificial intelligence has a major role to play not only in evolving all these clinical strategies for combating the disease but also for medical diagnostics based on radiologic imaging data. The most common application has been in the area of repurposing of existing drugs approved by the US-FDA for SARS-CoV-19 through various AI tools. Machine-learning models based on binding patterns of existing antiviral drugs with protein sequences from diverse virus species have been found to give fairly accurate predictions regarding the potential effectiveness of molecules against the proteomic sequence of the SARS-CoV-2 virus. Besides drug-treatment approaches, another powerful weapon against the virus is vaccines which normally take years to develop. Machine learning tools have been used to look for the immunogenic components of the virus, which could be exploited as good vaccine candidates. Thus, AI-driven modeling platforms are being used to accelerate drug design as well as vaccine design research for COVID-19.

11 Conclusions and future directions

Artificial intelligence has established itself as the foundation for research in all fields today and this includes its phenomenal contribution towards mitigating the challenges of the COVID-19 pandemic. This chapter has presented a review of AI applications in drug/vaccine development against the novel SARS-CoV-2 virus. AI has extensively contributed towards the understanding of the profile of the SARS-CoV-2-host protein interactions, which has been modeled on the lines of interactions of human cells with other viruses such as the Zika virus, Ebola virus, HIV virus, etc. Text- and data mining techniques have helped identify crucial proteins and molecular mechanisms that the SARS-CoV-2 virus has in common with previously known viruses. A huge list of molecular targets has built up from such studies in just a few months and is projected to uprise at an exponential rate as more studies are conducted. Prediction accuracy is critically dependent upon existing data and it should be understood that antiviral research is a particularly complex and ambiguous area of medicine. Even the mechanisms of action of existing drugs against various types of viruses are not unequivocal because the constantly evolving field of protein research will keep introducing newer protein targets with plausible newer mechanisms. The majority of the computational and biological studies have identified the M^{pro} as a crucial enzyme needed for the replication of coronavirus and having a huge potential for drug/vaccine targeting through de novo designing and repurposing.

By considering the timeline of novel drug development for establishing its efficacy and safety profile, a new molecule would take at least another 15–20 years before it reaches the patients. Hence, results from de novo design strategies don't seem attractive, at least for now, or till a suitable vaccine candidate arrives. Currently, repurposing appears to be a more attractive and time-effective approach due to the emergent requirement for an immediate cure for the disease. Several drugs predicted to be efficacious through AI or pharmacological screenings have been found to be useful in the clinical suppression of the disease. Repurposing strategies are generally based on homologies between molecular mechanisms or protein targets shared by the novel coronavirus and other viruses. Most of these studies may not be definitive in nature and could change in a future course. For example, the SARS-CoV-2 belongs to a different virus class from the HIV-1 virus and the homology between the two viruses is also disputed (Xiao et al., 2020). Inclusion of newer data will help design improvised AI models with predictions more accurately synced with clinical studies.

References

Abdel-Basset, M., Chang, V., Hawash, H., Chakrabortty, R. K., & Ryan, M. (2021). FSS-2019-nCov: A deep learning architecture for semi-supervised few-shot segmentation of COVID-19 infection. *Knowledge-Based Systems*, *212*, 106647.

Abdel-Basset, M., Chang, V., & Mohamed, R. (2020). HSMA_WOA: A hybrid novel slime mould algorithm with whale optimization algorithm for tackling the image segmentation problem of chest X-ray images. *Applied Soft Computing*, *95*, 106642.

Abdel-Basset, M., Chang, V., & Nabeeh, N. A. (2021). An intelligent framework using disruptive technologies for COVID-19 analysis. *Technological Forecasting and Social Change*, *163*, 120431.

Alcácerac, V., & Cruz-Machado, V. (2019). Scanning the industry 4.0: A literature review on technologies for manufacturing systems. *Engineering Science and Technology, An International Journal*, *22*, 899–919.

Beck, B. R., Shin, B., Choi, Y., Park, S., & Kang, K. (2020). Predicting commercially available antiviral drugs that may act on the novel coronavirus (2019-nCov), Wuhan, China through a drug-target interaction deep learning model. *bioRxiv*. https://doi.org/10.1101/2020.01.31.929547. 2020.01.31.929547.

Bung, N., Krishnan, S. R., Bulusu, G., & De Roy, A. (2020). Novo design of new chemical entities (NCEs) for SARS-CoV-2 using artificial intelligence. *ChemRxiv*. https://doi.org/10.26434/chemrxiv.11998347.v2. Preprint.

Cavasotto, C., & Filippo, J. D. (2020). *In silico* drug repurposing for COVID-19: Targeting SARS-CoV-2 proteins through docking and quantum mechanical scoring. *ChemRxiv*. https://doi.org/10.26434/chemrxiv.12110199.v2. Preprint.

Chen, Y. C., Lu, P. E., Chang, C. S., & Liu, T.-H. (2020). *A time-dependent SIR model for COVID-19 with undetectable infected persons*. http://arxiv.org/abs/2003.00122.

Cheng, F., Hong, H., Yang, S., & Wei, Y. (2017). Individualized network-based drug repositioning infrastructure for precision oncology in the panomics era. *Briefings in Bioinformatics*, *18*, 682–697.

Cheng, F., Murray, J. L., & Rubin, D. H. (2016). Drug repurposing: New treatments for zika virus infection? *Trends in Molecular Medicine*, *22*, 919–921.

Davenport, T., & Kalakota, R. (2019). The potential for artificial intelligence in healthcare. *Future Healthcare Journal*, *6*(2), 94–98.

Deep Mind. (2020). *Computational predictions of protein structures associated with COVID-19*. https://deepmind.com/research/open-source/computational-predictions-of-protein-structures-associated-with-COVID-19. Accessed August 04 2020.

Dhiman, G., Chang, V., Singh, K. K., & Shankar, A. (2021). ADOPT: Automatic deep learning and optimization-based approach for detection of novel coronavirus COVID-19 disease using X-ray images. *Journal of Biomolecular Structure & Dynamics*. https://doi.org/10.1080/07391102.2021.1875049.

Dubrow, A. (2020). *AI fast-tracks drug discovery to fight COVID-19*. https://www.tacc.utexas.edu/-/ai-fast-tracks-drug-discovery-to-fight-covid-19.

Durrant, J. D., Kochanek, S. E., Casalino, L., Ieong, P. U., Dommer, A. C., & Amaro, R. E. (2020). All-atom influenza virus simulations suggest new substrate binding mechanism. *ACS Central Science*, *6*, 189–196.

Dyall, J., Coleman, C. M., & Hart, B. J. (2014). Repurposing of clinically developed drugs for treatment of middle east respiratory syndrome coronavirus infection. *Antimicrobial Agents and Chemotherapy*, *58*, 4885–4893.

Epitopes World. (2020). *COVID-19 vaccine targets prediction*. https://epitopes.world/en.

Fast, E., Altman, R. B., & Chen, B. (2020). Potential T-cell and B-cell epitopes of 2019-nCoV. *bioRxiv*. https://doi.org/10.1101/2020.02.19.955484. 2020.02.19.955484.

Gaulton, A., Bellis, L. J., Bento, A. P., et al. (2012). ChEMBL: A large-scale bioactivity database for drug discovery. *Nucleic Acids Research*, *40*, D1100–D1107.

Guex, N., Peitsch, M. C., & Schwede, T. (2009). Automated comparative protein structure modeling with SWISS-MODEL and Swiss-PdbViewer: A historical perspective. *Electrophoresis*, *30*, S162–S173.

Gurucharan, M. K. (2020). *COVID-19: AI-enabled social distancing detector using OpenCV*. https://towardsdatascience.com/covid-19-ai-enabled-social-distancing-detector-using-opencv-ea2abd827d34.

Hoffmann, M., Kleine-Weber, H., Schroeder, S., et al. (2020). SARS-CoV-2 cell entry depends on ACE2 and TMPRSS2 and is blocked by a clinically proven protease inhibitor. *Cell*, *181*, 271–280.

HPC Wire StaffReport. (2020). *Quantum machine learning: A new weapon against COVID-19?*. https://www.hpcwire.com/2020/05/06/quantum-machine-learning-a-new-weapon-against-covid-19/.

Hu, F., Jiang, J., & Yin, P. (2020). *Prediction of potential commercially inhibitors against SARS-CoV-2 by multi-task deep model*. arXiv:2003.00728v2 [q-bio.MN]. Accessed August 26, 2021.

Jaimes, J. A., André, N. M., Chappie, J. S., Millet, J. K., & Whittake, G. R. (2020). Phylogenetic analysis and structural modeling of SARS-CoV-2 spike protein reveals an evolutionary distinct and proteolytically sensitive activation loop. *Journal of Molecular Biology*, *432*, 3309–3325.

Jin, C., Chen, W., Cao, Y., Xu, Z., Zhang, X., & Deng, L. (2020). *Development and evaluation of an AI system for COVID-19 diagnosis* (p. 1e23). https://doi.org/10.1101/2020.03.20.20039834.

Johansen, L. M., Brannan, J. M., Delos, S. E., et al. (2013). FDA-approved selective estrogen receptor modulators inhibit ebola virus infection. *Science Translational Medicine*, *5*, 190ra179.

Ke, Y., Peng, T., Yeh, T., et al. (2020). Artificial intelligence approach fighting COVID-19 with repurposing drugs. *Biomedical Journal*, *43*, 355–362.

Kim, J. D., Ohta, T., Tateisi, Y., & Tsujii, J. (2003). GENIA corpus–semantically annotated corpus for bio-textmining. *Bioinformatics*, *19*, 180–182.

Koppal, T. (2020). *Artificial intelligence: Applications in the pandemic and beyond*. https://www.biocompare.com/Editorial-Articles/565713-Artificial-Intelligence-Applications-in-The-Pandemic-and-Beyond/.

Kumar, S. (2020). *COVID-19: A drug repurposing and biomarker identification by using comprehensive gene-disease associations through protein-protein interaction network analysis*. Preprints 2020030440. https://doi.org/10.20944/preprints202003.0440.v1.

Kumar, A., Gupta, P. K., & Srivastava, A. (2020). A review of modern technologies for tackling COVID-19 pandemic. *Diabetes and Metabolic Syndrome: Clinical Research & Reviews*, *14*, 569e573.

Kurji, N. (2020). *China's institute of materia medica partners with cyclica on innovative drug repurposing for COVID-19*. https://www.cio.com/article/3292616/how-singapore-is-using-artificial-intelligence.html.

Li, Y., Zhang, J. G., Goncalves, J., et al. (2020). An interpretable mortality prediction model for COVID-19 patients. *Nature Machine Intelligence*, *2*, 283–288.

Li, Y., Zhang, J., Wang, N., et al. (2020). Therapeutic drugs targeting 2019-ncov main protease by high-throughput screening. *bioRxiv*. https://doi.org/10.1101/2020.01.28.922922. 2020.01.28.922922.

Lin, L., & Hou, Z. (2020). Combat COVID-19 with artificial intelligence and big data. *Journal of Travel Medicine*, 1–4. https://doi.org/10.1093/jtm/taaa080.

Liu, X., & Wang, X. (2020). Potential inhibitors against 2019-ncov coronavirus m protease from clinically approved medicines. *Journal of Genetics and Genomics*, *47*(2), 119–121.

Lu, J., & Sun, P. D. (2020). High affinity binding of SARS-CoV-2 spike protein enhances ACE2 carboxypeptidase activity. *The Journal of Biological Chemistry*. https://doi.org/10.1074/jbc.RA120.015303.

Lu, R., Zhao, X., Li, J., et al. (2020). Genomic characterisation and epidemiology of 2019 novel coronavirus: Implications for virus origins and receptor binding. *Lancet*, *395*, 565–574.

Madrid, P. B., Chopra, S., Manger, I. D., et al. (2013). A systematic screen of FDA-approved drugs for inhibitors of biological threat agents. *PLoS One*, *8*, e60579.

Magar, R., Yadav, P., & Farimani, A. B. (2021). Potential neutralizing antibodies discovered for novel corona virus using machine learning. *Scientific Reports*, *11*(1), 5261.

Malone, B., Simovski, B., Moliné, C., et al. (2020). Artificial intelligence predicts the immunogenic landscape of SARS-Cov-2 leading to universal blueprints for vaccine designs. *Scientific Reports*, *10*, 22375.

Mani, D., Wadhwani, A., & Krishnamurthy, P. T. (2019). Drug repurposing in antiviral research: A current scenario. *Journal of Young Pharmacists*, *11*, 117–121.

Maslan, C. (2020). *Social distancing detection for COVID-19*. https://blog.camio.com/social-distancing-detection-for-covid-19-a5c9e8d6c760.

Medicircle. (2020). *Innoplexus announces major findings in potential combination therapies for COVID-19*. https://www.innoplexus.com/news/innoplexus-announces-major-findings-in-potential-combination-therapies-for-covid-19/. Accessed August 04, 2020.

Metabiota Epidemic Tracker. (2020). https://www.epidemictracker.com. Accessed August 04, 2020.

Narayana, D., Arpna, D., Peeyushi, S., Samiksha, S., Sanjay, V., & Sneha, P. (2020). *Artificial intelligence as COVID-19 drug discovery booster—Express healthcare*. https://mc.ai/artificial-intelligence-as-covid-19-drug-discovery-booster-express-healthcare/.

Narin, A., Kaya, C., & Pamuk, Z. (2021). Automatic detection of coronavirus disease (COVID-19) using X-ray images and deep convolutional neural networks. *Pattern Analysis and Applications*, *24*, 1207–1220.

Naudé, W. (2020). *Artificial Intelligence vs COVID-19: Limitations, constraints and pitfalls*. https://doi.org/10.1007/s00146-020-00978-0.

Nguyen, T. T., Abdelrazek, M., Nguyen, D. T., Aryal, S., Nguyen, D. T., & Khatami, A. (2020). Origin of novel coronavirus (COVID-19): A computational biology study using artificial intelligence. *bioRxiv*. https://doi.org/10.1101/2020.05.12.091397. 2020.05.12.091397.

Nguyen, D. D., Gao, K., & Chen, J. (2020). Potentially highly potent drugs for 2019-nCoV. Preprint *bioRxiv*. https://doi.org/10.1101/2020.02.05.936013. 2020.02.05.936013.

Nguyen, T. T., Waurn, G., & Campus, P. (2020). *Artificial intelligence in the battle against coronavirus (COVID-19): A survey and future research directions*. https://doi.org/10.13140/RG.2.2.36491.23846.

Nukoolkarn, V., Lee, V. S., & Malaisree, M. (2008). Molecular dynamic simulations analysis of ritonavir and lopinavir as SARS-CoV 3CL(pro) inhibitors. *Journal of Theoretical Biology*, *254*, 861–867.

Ong, E., Wang, H., & Wong, M. U. (2020). Vaxign-ML: Supervised machine learning reverse vaccinology model for improved prediction of bacterial protective antigens. *Bioinformatics*, *36*(10), 3185–3191.

Ong, E., Wong, M. U., & Huffman, A. (2020). COVID-19 coronavirus vaccine design using reverse vaccinology and machine learning. Preprint *bioRxiv*. https://doi.org/10.1101/2020.03.20.000141. 2020.03.20.000141. 2020.

Oztemel, E., & Gursev, S. (2020). Literature review of industry 4.0 and related technologies. *Journal of Intelligent Manufacturing*, *31*(4), 127–182.

Popova, M., Isayev, O., & Tropsha, A. (2018). Deep reinforcement learning for de novo drug design. *Science Advances*, *4*(7), eaap7885.

Rao, A. S. R. S., & Vazquez, J. A. (2020). Identification of COVID-19 can be quicker through artificial intelligence framework using a mobile phone-based survey in the populations when cities/towns are under quarantine. *Infection Control and Hospital Epidemiology*, *2020*, 1–5. https://doi.org/10.1017/ice.2020.61.

Richardson, P., Corbellino, M., & Stebbing, J. (2020). Baricitinib for COVID-19: A suitable treatment? authors' reply. *The Lancet Infectious Diseases*, *2*(393), 52e65.

Richardson, P., Griffin, I., & Tucker, C. (2020). Baricitinib as potential treatment for 2019-ncov acute respiratory disease. *Lancet*, *395*, E30–E31.

Rosenblatt, L. (2020). *New research institute puts $5 million toward AI to battle COVID-19 spread and its social impact*. https://www.post-gazette.com/business/tech-news/2020/06/23/artificial-intelligence-COVID19.

Santos, J., Brierley, S., Gandhi, M. J., Cohen, M. A., Moschella, P. C., & Declan, A. B. L. (2020). Repurposing therapeutics for potential treatment of SARS-CoV-2: A review. *Viruses*, *12*(7), 705.

Senior, A. W., Evans, R., & Jumper, J. (2020). Improved protein structure prediction using potentials from deep learning. *Nature*, *577*, 706–710.

Shu, Y., & McCauley, J. (2017). GISAID: Global initiative on sharing all influenza data—From vision to reality. *Euro Surveillance*, *22*(13), 30494.

Skelton, D. J., Alsobhe, A., & Anastasi, E. (2020). *Drug repurposing prediction for COVID-19 using probabilistic networks and crowdsourced curation*. arXiv:2005.11088 [q-bio.MN].

Smartcity. (2020). *Can AI help the world find the medicine to treat COVID-19?*. https://www.smartcity.press/ai-accelerating-covid-vaccine/.

Sodani, P., Mucci, L., & Girolimetti, R. (2020). Successful recovery from COVID-19 pneumonia after receiving baricitinib, tocilizumab, and remdesivir. A case report: Review of treatments and clinical role of computed tomography analysis. Version 2. *Respiratory Medicine Case Reports*, *31*, 101115.

Sohrabi, C., Alsafi, Z., & O'Neill, N. (2020). World Health Organization declares global emergency: A review of the 2019 novel coronavirus (COVID-19). *International Journal of Surgery*, *76*, 71–76.
Stebbing, J., Phelan, A., & Grifn, I. (2020). COVID-19: Combining antiviral and anti-infammatory treatments. *Lancet*, *20*, 400–401.
Stokes, J. M., Yang, K., Swanson, K., Jin, W., Cubillos-Ruiz, A., & Donghia, N. M. (2020). A deep learning approach to antibiotic discovery. *Cell*, *180*, 688–702.
Tang, B., He, F., Liu, D., Fang, M., Wu, Z., & Xu, D. (2020). AI-aided design of novel targeted covalent inhibitors against SARS-CoV-2. *BioRxiv*. https://doi.org/10.1101/2020.03.03.972133. Preprint.
Tarasova, O., Ivanov, S., Filimonov, D. A., & Poroikov, V. (2002). Data and text mining help identify key proteins involved in the molecular mechanisms shared by SARS-CoV-2 and HIV-1. *Molecules*, *25*, 2944–2958.
The Indian Practitioner. (2020). https://theindianpractitioner.com/2020/07/08/iiit-delhi-develops-artificial-intelligence-model-to-repurpose-existing-drugs-for-treating-covid-19/. Accessed August 02, 2020.
The Pharma Letter. (2020). https://www.thepharmaletter.com/article/who-launches-new-technology-access-initiative-for-covid-19. Accessed August 04, 2020.
The Science Advisory Board. (2020). *Scripps research expands drug repurposing efforts for COVID-19*. https://www.scienceboard.net/index.aspx?sec=sup&sub=cell&pag=dis&ItemID=665.
The Science Advisory Board staff. (2020). https://www.scienceboard.net/index.aspx?sec=sup&sub=BioChem&pag=dis&ItemID=587. Accessed September 09, 2020.
Ting, D. Z. W., Carin, L., Dzau, V., & Wong, T. Y. (2020). Digital technology and COVID-19. *Nature Medicine*, *26*, 458–464.
Tuli, S., Tuli, S., Tuli, R., & Gill, S. S. (2020). Predicting the growth and trend of COVID-19 pandemic using machine learning and cloud computing. *Internet of Things*, *11*, 100222.
Vaishya, R., Javaid, M., Khan, I. H., & Haleem, A. (2002). Artificial intelligence (AI) applications for COVID-19 pandemic. *Diabetes and Metabolic Syndrome: Clinical Research and Reviews*, *14*, 337e339.
Vashisht, A., Bagaria, A., Kumari, S., Singh, O. K., Gupta, A., & Bhagat, N. (2020). Artificial intelligence & Covid- 19—A review. *International Journal of Research in Health and Allied Sciences*, *6*(3), 14–17.
Verma, M., & Bansal, D. (2020). Novel potential inhibitors against SARS-Cov-2 using artificial intelligence. *ChemRxiv*. https://doi.org/10.26434/chemrxiv.12228362.v4. Preprint.
Wang, S., Kang, B., Ma, J., et al. (2020). *A deep learning algorithm using CT images to screen for corona virus disease (COVID-19)*. 1e28, medRxiv 2020.02.14.20023028.
Wang, S., Zha, Y., Li, W., et al. (2020). A fully automatic deep learning system for COVID-19 diagnostic and prognostic analysis. *The European Respiratory Journal*. https://doi.org/10.1183/13993003.00775-2020. In press.
Weininger, D., Weininger, A., & Weininger, J. L. (1989). Smiles. 2. Algorithm for generation of unique smiles notation. *Journal of Chemical Information and Computer Sciences*, *29*, 97–101.
Worldometer. (2020). https://www.worldometers.info/coronavirus/?utm_campaign=homeAdvegas1. Accessed August 18, 2020.
Xiao, C., Li, X., Liu, S., Sang, Y., Gao, S., & Gao, F. (2020). HIV-1 did not contribute to the 2019-nCoV genome. *Emerging Microbes & Infections*, *9*(1), 378–381.
Xu, Z., Peng, C., Shi, Y., et al. (2020). Nelfinavir was predicted to be a potential inhibitor of 2019-ncov main protease by an integrative approach combining homology modelling, molecular docking and binding free energy calculation. *bioRxiv*. https://doi.org/10.1101/2020.01.27.921627.

Zhang, H., Saravanan, K. M., & Yang, Y. (2020a). Deep learning based drug screening for novel coronavirus 2019-nCov. *Interdisciplinary Sciences, Computational Life Sciences*. https://doi.org/10.1007/s12539-020-00376-6.

Zhang, H., Saravanan, K. M., & Yang, Y. (2020c). Deep learning based drug screening for novel coronavirus 2019-nCov. *Interdisciplinary Sciences, Computational Life Sciences*. https://doi.org/10.1007/s12539-020-00376-6.

Yu, F., & Koltun, V. (2016). Multi-scale context aggregation by dilated convolutions. *International Conference on Learning Representations*. arXiv:1511.07122v3 [cs.CV].

Zhang, C., Wu, Z., Li, J., Zhao, H., & Wang, G. (2020). Cytokine release syndrome in severe COVID-19: Interleukin-6 receptor antagonist tocilizumab may be the key to reduce mortality. *International Journal of Antimicrobial Agents*, *55*(5), 105954.

Zhao, Y., Zhao, Z., Wang, Y., Zhou, Y., Ma, Y., & Zuo, W. (2020). Single-cell RNA expression profiling of ACE2, the receptor of SARS-Cov-2. *bioRxiv*. https://doi.org/10.1101/2020.01.26.919985. 2020.01.26.919985.

Zhavoronkov, A. (2018). Artificial intelligence for drug discovery, biomarker development, and generation of novel chemistry. *Molecular Pharmaceutics*, *15*(10), 4311–4313.

Zhavoronkov, A., Aladinskiy, V., Zhebrak, A., et al. (2020). Potential COVID-2019 3C-like protease inhibitors designed using generative deep learning approaches. *ChemRxiv*. https://doi.org/10.26434/chemrxiv.11829102.v2. Preprint.

Zhou, Y., Hou, Y., & Shen, J. (2020). Network-based drug repurposing for novel coronavirus 2019-nCoV/SARS-CoV-2. *Cell Discovery*, *6*(14), 2020.

Zhou, B., Yang, Z., Feng, Q., et al. (2017). Aurantiamide acetate from baphicacanthus cusia root exhibits anti-inflammatory and antiviral effects *via* inhibition of the NF-κB signaling pathway in influenza a virus-infected cells. *Journal of Ethnopharmacology*, *199*, 60–67.

Zhou, P., Yang, X., & Wang, X. (2020). A pneumonia outbreak associated with a new coronavirus of probable bat origin. *Nature*, *579*, 270–273.

CHAPTER 9

The economic impact of covid-19 and the role of AI

Chandra Bhanu Nayak, Prasant Kumar Nanda, Snigdha Tripathy, Sukanta Chandra Swain, Chinmay Kumar Das, and Rojalin Sahu
Kalinga Institute of Industrial Technology, Deemed to be University, Bhubaneswar, India

1 Introduction

As the pandemic unleashed by SERS-COVID-19 tramples down lives and livelihood, economists worldwide ask a very crucial question, i.e. how the business will be in a post-COVID world (Rohatgi, 2020). May be with industries shutting down with the widespread loss of jobs, we are looking down the barrel (Javaid et al., 2020). No doubt there is going to be a wide chasm between the haves and the have-nots. GDPs across the world are cascading down like a house of cards. This can very well be understood if we study the scenario in the world's most powerful country i.e. the United States. Analysts say that it may take years to recover the status quo. More than 25% of the working-age population of the United States have already lost their regular jobs and are applying for government subsistence. The scenario in Europe, Asia, and Pacific and Africa is no better. With no potential cure available shortly, sociologists apprehend that we may have to live with the pandemic (Garnier, Sandvik, & Cellini, 2020; Harari, 2020; Mason, 2020; Taskinsoy, 2020). Though lockdowns and social distancing have made a dent in the spread of the virus, yet we see that we are moving towards community transmission. In India, we have a huge population and an ever-widening gap between the few rich people and the majority of the population living on the brink (Nguyen et al., 2020). Artificial Intelligence has numerous packages in today's society. It is vital for today's time because it can clear up complicated issues in a green manner in a couple of industries, including healthcare, entertainment, finance, education, etc. AI is making our each day existence greater cushy and fast (Figs. 1 and 2). Artificial intelligence can rework job tasks and occupations. This column uses US online job postings throughout 2010–19 to indicate how absolute and the relative demand for AI-related skills has matured across all business sectors and

Novel AI and Data Science Advancements for Sustainability in the Era of COVID-19
https://doi.org/10.1016/B978-0-323-90054-6.00002-7

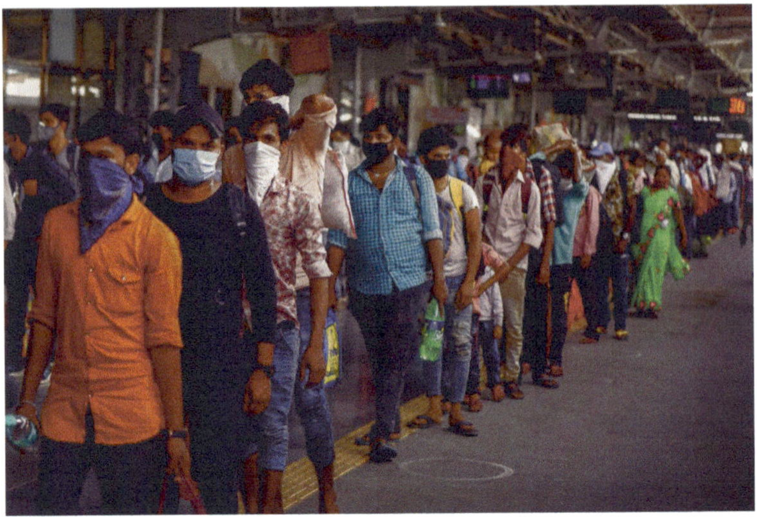

Fig. 1 Migrant labors during pandemic. *(Source: Original.)*

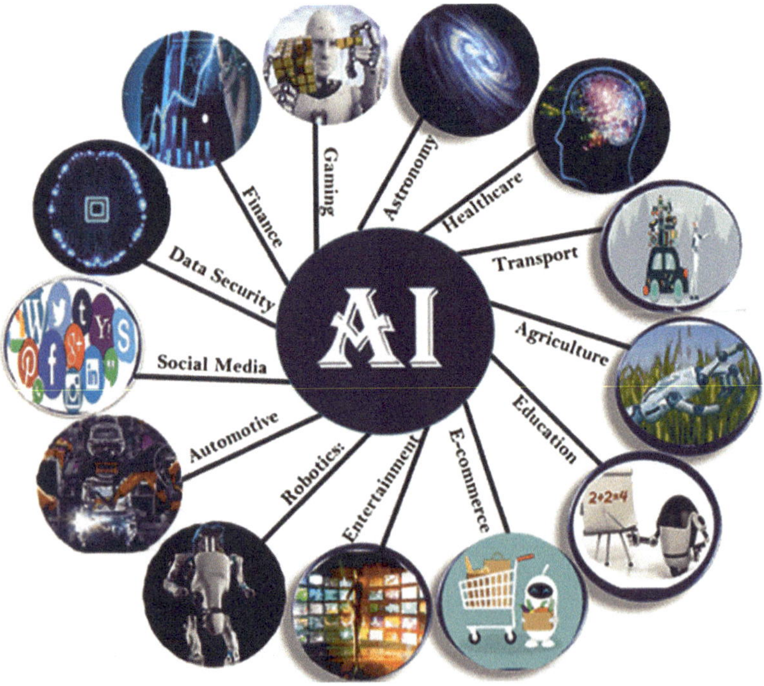

Fig. 2 AI utility. *(Source: https://www.javatpoint.com/application-of-ai.)*

occupation groups. According to this record, with the aid of using Canadian-primarily based start-up totally, 400 international human beings have appeared as pinnacle AI skills. Among those AI experts, 50% belonged to the United States, with some of 10.295, observed using China, which had 2525 AI experts. The record, in addition, said that rank three and four are for the United Kingdom, Germany and Canada with some of 1457, 935, and 815 respectively. According to the record, India stands at quantity nine in those AI experts standing, with the quantity of AI experts being 555 (Fig. 3). This pandemic is converting a pre-existing economic slowdown into a full-fledged crisis. This may cause lasting damage to the already weakened economy. On the economic front, India was in pretty bad shape as Covid-19 struck. Just before the outbreak of the pandemic, we had seen that economic growth was slumping. The investment was not coming and unemployment, as well as poverty, were rising. A clogged financial sector choked the economy even as battered public finances were of limited help. In the wake of this situation, we have witnessed reverse migration of lakhs of migrant labor who walked home listlessly, some dying on the way as the entire economy came to a screeching halt. As we watched this humanitarian disaster helplessly, it is a sad story that these people have contributed to the spread of the disease. This has created nationwide indignation. Even as the

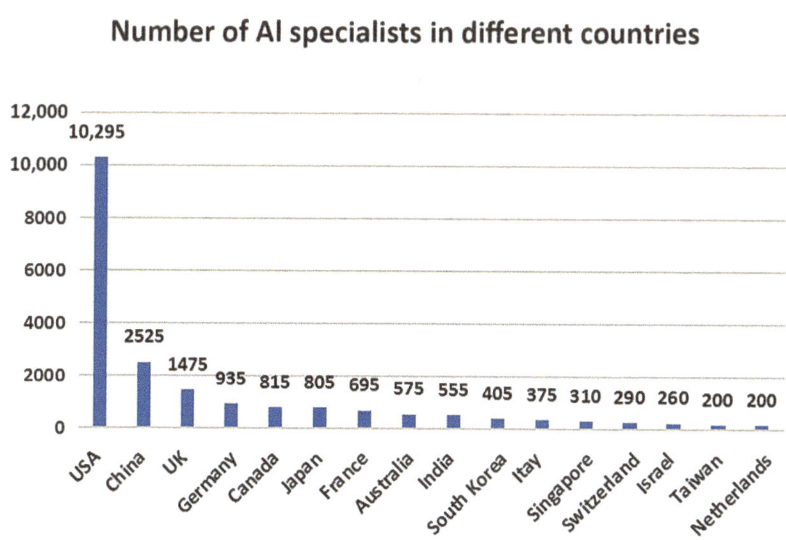

Fig. 3 World ranking of countries with number of AI scholars. *(Source: https://analyticsindiamag.com/india-among-top-9-in-worldwide-rankings-in-terms-of-ai-specialists/.)*

government has started Unlock I and II, to unleash economic activity and save livelihoods, yet it is presumed that it may take years to nurse back the economy to the much-hyped 5 trillion-dollar economy. Each and every sector of the economy around the world has been badly affected, which has triggered a recession-like situation. Traditional organized and unorganized sectors have shredded jobs. Further, there is every likelihood that the middle and lower-middle-class will be pushed into poverty. The pandemic threatens millions of people and has challenged the country's already overburdened health system like never before. Limited state capacity has severely affected the Covid-19 response (Stephany et al., 2020).

If we take a careful look at the banking system, which forms the backbone of the country's economy, we find that the balance sheets have taken a hit. Bad loans or NPAs are rising in the banks. The RBI has reduced the repo rate to push credit into the shagging sectors of the economy (Dev & Sengupta, 2020; Stephany et al., 2020). The government has started multiple schemes like "Atma Nirvar Bharat" or self-reliance to kick start the economy (Singh, Javaid, Haleem, & Suman, 2020). The government wants the resident Indians to boost local products. "Vocal for local" is the call for the day (Ahluwalia, Mahto, & Guerrero, 2020). The national government thinks that giving a boost to the MSME sector will give the necessary impetus to turn around the clock. The world over the pandemic has pushed back the economy by decades. The largest and most powerful economy i.e., the United States, is in the doldrums. Even China, from where the pandemic originated, has suffered. Never before has humanity been put to such a grinding test. World over the rich and the poor, old and the young, sick and healthy is facing the same music. Four out of five workers worldwide have lost their jobs. The global economy has shrunk by 20% and India's by over 30%. Over 80% have lost their livelihood. From New York to New Delhi, people are crying and imploring their governments to provide minimum subsistence levels for the people. ILO has said that workers and businesses are facing a catastrophe of unimaginable magnitude in both developed and developing countries. Things have to move fast in a decisive manner to set things right. The world has come to a stage where we have to choose between survival and collapse (Ozili & Arun, 2020).

Presently the beleaguered economy is trying to limp back to some kind of normalcy. It is good news that the Government of India has introduced Rs 20 lakh crore booster packages to kick start the economy (Singh & Gupta, 2010). The social distortions and hardships imposed by the pandemic continue to cast their pall of gloom. We have realized that the nation cannot remain in continued lockdown as its economic fallout is quite debilitating. The long-lasting consequences of the pandemic and consequent restrictions present a gloomy future with repeated disruptions, pain and loss. Coronavirus has caused great anguish to daily wage earners, street vendors, and small shop owners, contract employees and debt-ridden farmers. Under this background, we have chosen the topic for research keeping the aim in mind that how artificially intelligent data can play a vital role in addressing these problems (McCall, 2020). The topic for research will be a pioneering work and will provide the impetus for future social scientists to take forward the activity. This is important because our total society is facing the problem and fighting for survival. We understand that there is no past attempt to solve the research problem and answer the research questions. Under these circumstances, the research paper will attempt to provide tangible solutions using artificial intelligence, especially migrant workers. It is a fact that the solution lies in upskilling the workers with artificial intelligence tools like learning techniques, soft computing, expert systems, computational intelligence, data mining, biometrics etc. We propose to adopt a comparative study in analyzing the pre-Covid era and the post-Covid era. We will study in-depth the psychological response of the migrant laborers. Literature study will also be taken up to study similar situations in the past. Over the last many centuries, this world has been affected by so many pandemics. But there is a basic difference between the past and the present. Today because of technological advancement, we have effective tools to fight off any emergencies. Scientists are toiling day and night to prepare an effective vaccine for the elusive virus all over the world. Analysis of past data will lead the way to suggest a road map to recover from this situation. Restoring the livelihood of millions, providing online platforms for continuing education and strengthening health care systems are the issues that need to be addressed (Hyder et al., 2020; Laghi, 2020; Li et al., 2020; Lin & Hou, 2020; Nguyen et al., 2020; Rao & Vazquez, 2020; Singh et al., 2020; Yang, Gentile, Shen, & Cheng, 2020). The study will use methodologies like observation, survey, primary/secondary data analysis and archival data mining to study the topic comprehensively. Sample studies will be carried out to facilitate arriving at the desired objectives.

This chapter is an attempt to answer the research questions as follows;

- Whether the economic system in India was not prepared to handle a grave crisis of this magnitude?
- What is the economic fallout of the pandemic-qualitatively and quantitatively?

- Whether this will trigger a chain reaction for the new world order?
- Whether humankind can use the tools of Artificial Intelligence to predict and visualize this type of pandemic in the future.

AI may "help with the next pandemic." In the meantime, gathering extensive diagnostic data on who is infectious will be essential to save lives and limit economic damages (Baldwin & Tomiura, 2020; Bloom et al., 2020; Dewatripont et al., 2020).

2 Economic impact of Covid-19

The economic impact of Covid-19 is endemic. The global effect can be ascertained from the fact that major banks in the United States have set aside nearly 28 billion dollars this quarter for potential bad loans. Governments across the world have unveiled plans in the form of deferments, direct assistance and unemployment aid. But the fact is that this massive stimulus may not be enough to support society in the long run. There is little likelihood that the economy will bounce back shortly. Table 1 presents the projection of real GDP growth (in percentage, basis: year-on-year) done by the Organization for Economic Cooperation and Development (OECD). Colors in the table indicate the direction of revisions since the December, 2020 Economic Outlook.

Post the first wave of Covid-19, activities in many sectors have picked up and partly adapted to pandemic restraints. Government stimulus coupled with vaccine roll-out, which is gaining momentum, is presumptive to furnish a stellar boost to economic activities in a slow and uneven manner. However, outlooks for sustainable growth differ largely between countries and sectors. Quicker and more efficacious vaccination preparation across the globe is captious. Prospects have been reinforced over recent months, with signs of a movement in merchandize trade and industrial production becoming unsubtle by the end of 2020. Global GDP growth is now projected to be 5.6% this year, i.e., in 2021, an upward revision of more than one percentage point from the December *OECD Economic Outlook*. World output is expected to reach pre-pandemic levels by mid-2021, but much will depend on the race between vaccines and emerging virus variants. The global vaccine roll-out remains uneven, with restrictions remaining in some countries and sectors. The outlook for growth would improve (upside scenario) if the production and distribution of doses accelerate, is better coordinated worldwide, and gets ahead of virus mutations. This would allow containment

Table 1 Projections of real GDP growth (%, *year-on-year, colors indicate the direction of revisions since the December 2020 Economic Outlook*).

	2020	2021	2022
World	−3.4	5.6	4.0
Australia	−2.5	4.5	3.1
Canada	−5.4	4.7	4.0
Euro area	−6.8	3.9	3.8
Germany	−5.3	3.0	3.7
France	−8.2	5.9	3.8
Italy	−8.9	4.1	4.0
Spain	−11.0	5.7	4.8
Japan	−4.8	2.7	1.8
Korea	−1.0	3.3	3.1
United Kingdom	−9.9	5.1	4.7
United States	−3.5	6.5	4.0
G20	−3.2	6.2	4.1
Argentina	−10.5	4.6	2.1
Brazil	−4.4	3.7	2.7
China	2.3	7.8	4.9
India	−7.4	12.6	5.4
Indonesia	−2.1	4.9	5.4
Mexico	−8.5	4.5	3.0
Russia	−3.6	2.7	2.6
Saudi Arabia	−4.0	2.6	3.9
South Africa	−7.2	3.0	2.0
Turkey	1.8	5.9	3.0

Note:

(orange)	Downward revision, by 0.3 pp or more
(grey)	No change or smaller than 0.3 pp
(green)	Upward revision, by 0.3 pp or more

Source: OECD (2021). OECD Economic Outlook, interim report March 2021.

measures to be relaxed more rapidly and global output to approach pre-pandemic projections for activities. But consumption expenditure and business certainty would be striking (downside scenario) if vaccination programs are not accelerated enough to incise infection rates or if new variants become more far-flung and necessitate alterations to incumbent vaccines. However, considering the intensity of cruelty of the second wave of Covid-19, the projections done may get disturbed.

It needs less to mention that at present, a big cloud of uncertainty surrounds the global outlook. The future will depend on how the nations manage health crisis, including whether the vaccines successfully counter the new COVID-19 strains/ genetic mutations or they prolong the pandemic;

the effectiveness of policy formulations to limit continuous damage to the economy; the emerging market dynamics and fiscal/administrative measures to reign in escalating prices; and the flexibility of the economy to bounce back.

While the pandemic rages without a break, national policies should first focus on minimizing the long-term effect of the crisis, increasing health care spending as a percentage of GDP, providing well-targeted financial support to crucial sectors, and maintaining accommodative monetary policy while monitoring financial stability risks. Then, as the recovery progresses, policymakers will need to limit long-term economic damage by more production (for example, public investment) and increasing incentives for an efficient allocation of productive resources. It is a delicate balance, especially given the prevailing uncertainty as needed). Simultaneously, long-term challenges like boosting productivity, improving policy frameworks, and addressing climate change issues cannot be sidetracked. Widespread mass vaccination covering the frontline workers, vulnerable groups of the population, working-age group citizens hold the key to early recovery.

Strong international cooperation is vital for achieving these objectives and ensuring that emerging market economies and low-income developing countries continue to narrow the gap between their living standards and those of high-income countries.

- On the health care front, this means ensuring adequate worldwide vaccine production and universal distribution at affordable prices—including through sufficient funding for the vaccination facility—so that all countries can quickly and decisively beat back the pandemic.
- Vaccination should be universal and suitable tax incentives can be given.
- Moreover, strong cooperation is needed to resolve economic issues.

Global trade is estimated to have fallen by 5.3% in 2020 but is projected to grow by 8.0% in 2021. Major advanced economies, which comprise 60% of global economic activity, are projected to operate below their potential output level through at least 2024. A resurgence in infectious cases in Europe, the United States, Japan, Brazil, India, and various developing economies renewed calls for lockdowns and curfews and threatened to weaken or delay a potential sustained economic recovery into mid to late 2021.

In India, the situation is no better (The Economics Times, 28 Apr 2021). It is a very precarious situation in India, home to nearly 140 crore human beings i.e., around 20% of the global population. The onset of the second wave has dashed the hope of a V-shaped economic recovery as the pandemic rages around with lives and livelihoods being lost like ninepins.

Presently the Indian economy, the world's fifth-largest, is under the long shadow of long COVID as the aftereffect of the long seize is beginning to have its deleterious effect on the economy. Economists and rating agencies have downgraded India's growth. We can hope for a maximum nine plus % GDP growth against an IMF forecast of 11.5% as of January 2021.

If we look at Budget 2020, it estimated the country's GDP at around Rs 203 lakh crore or $2.8 trillion. The economy was projected to grow by 10% and reach approximately Rs 220 lakh crore. On the other hand, the economy contracted around 8%, downgrading to Rs 195 crore or around $2.6 trillion, pegging at the figures of 2018. This year for 2021–22, as per the budget, the economy was projected to touch Rs 222 lakh crore or $3.00 trillion. But as of now, it seems that the economy may make it to the pre-pandemic level of 2019 by early 2022.

It was expected that the budget of 2021 would achieve its targets because of the ongoing vaccination program, privatization of selected public sectors to fund expenditure and boosting the financial sectors. The assumption was that this would set the cycle rolling for demand creation, jobs, income generation, consumption, investment opportunities and resultant growth.

But as things turn out today, there is little movement on the economic front. The second wave of the muted virus rages across the length and breadth of the country in rural and urban India. As a result of this, the nation as a whole is under lockdown and shut down. Spiraling caseloads in lakhs and increasing mortality rate have put brakes on livelihoods. The retail and MSME sectors are the worst hit. The retail sector contributes 10% to the GDP and creates jobs. Now, most of the employees have lost their jobs or are irregularly paid. In this scenario, the funding roadmap via privatization may not be off-take as this is not the best time disinvesting. Lately, 90% of the economic activity, movement of goods and services are under standstill effect. Epidemiologists and economists are accessing that the Virus curve may flatten by July end and the economic unlock process will begin.

The Indian economy is on the ventilator and struggling to rebound back. The human distress is clearly visible with prices of diesel, petrol almost increasing daily and prices of essential commodities skyrocketing through the roof. Gross mismanagement of the fragile health care system is clearly showing up with an acute shortage of medical oxygen, lifesaving drugs, Poor infrastructure, ICU beds, ventilators etc.

Though the situation looks pretty gloomy, economic pundits hope things may turn better as vaccination speeds up and the curve flattens. The timeline looks hazy as of now.

A whopping 6.2% of the country's GDP, or about Rs 12.5 lakh crore, is required to fight the Covid-19 pandemic in India with intensified measures. In India, the Union health budget is 1.3% of the GDP, and the Union Govt. has allocated Rs 15,000 crore only towards the public health crisis so far. As a result, the cost of COVID-19 care will be 5 times the annual allocation on health and 75 times the fund allocated to fight till now. Presently we are witnessing a recession-like situation around the globe. But it is no ordinary, normal situation. It is estimated that India's economy will contract in 2020–21 from −5% to −15%. The GDP has hit rock bottom in April and May 2020. Businesses are struggling to survive. As an after-the-shoot, we are witnessing firms downsizing employees and reducing salaries to bridge the gap between revenue generation and recurring expenditure. We are gradually likely to see more and more outlets in the retail or FMCG sectors, whether automobile, dress, or eateries, to reduce by 30% due to lack of demand generation. This will result in jobs lost, business space vacated, less power consumption and elimination of other needs. This will, in most likelihood, usher in an era of stagflation, i.e., inflation will rise with a high unemployment rate and subdued demand. In addition to this, we are also witnessing a nonstop rise in fuel prices and their effect on ancillary sectors. Economists fear that there may be a situation where there will be no demand in our country, but prices will grow up continuously. Production units will be unable to cut costs due to the reduced no. of units. Social scientists believe that people will spend less and hold on to their liquidity or cash due to reduced income and uncertainty about the future until normalcy is restored in normal economic activity. Of course, nobody will invest in foreign travel or branded apparel, or consumer goods shortly. With each and every sector of the economy being affected and coupled with the migrant labor crisis and other crises, the middle class, which forms the backbone of the economy, is already pushed into economic stress and poverty. The migrant labor class, which forms the majority of the workforce in factories in distant towns, had left the cities for their native villages at the height of the crisis. Now they are slowly returning to their previous workplaces in the absence of any employment opportunity in their native places. But can all these people get back their jobs? In the absence of any other alternative, the farm or agriculture sector is the only avenue that is expected to do well. Then we may see another flow of reverse migration. The banking sector constitutes an integral part of the financial sector. The RBI revealed in its Financial Stability Report (FSR) that the economic stress induced by Covid-19 could push gross bad loans in banks to their highest in two decades. The loan moratorium has suppressed the real stress on Indian Banks. The report warns that in this very stressed scenario, the gross NPA of the banking sector would rise to as high as 14.7% of total loans as of March 2021. It predicts a very gloomy picture for the Indian banking sector, at least in the immediate future. The pandemic has also added to the volatility of the financial markets. The economy has already shrunk to abysmal depth. It is a million-dollar question, how soon we can see a recovery.

It is not that the pandemic has taken a toll in India, but it has done so globally. From where the pandemic came from, China is the first major country with economic growth falling by 23%. But significantly, the case counts to be 81 thousand only out of a population of 138 crores. Governments across the world are providing fiscal stimulus packages (Morey, 2004). Globally it is difficult to predict the long-term effects. But sectors like tourism, hospitality, public events are under a lot of stress. On the contrary, sectors like retail, logistics, education, manufacturing, health care, media, telecom, banking, and insurance face radical changes and new opportunities. The financial sector is facing long-term challenges and defaults are rising. Covid-19 has disrupted these industries, but the business models will be more or less dependent on increased digitization.

AI tools like automation, cloud computing, cybersecurity, the Internet of things and data analytics are the need of the hour. Health care industries will have to depend upon data analytics tools to track their business progress and track their investments. In the long term, real-time data will be required to monitor the health conditions. Manufacturing units associated with health care will have to produce personal protective equipment and healthcare technology. Retailers will have to go for large-scale digitalization, considering that consumers will not physically come to the stores. Supply chains and logistics firms will have to be more sensitive to the serious pandemic situation. Online education using new AI technology has already become the order of the day instead of the traditional education systems in schools and colleges (Bragazzi et al., 2020; Luckin, Holmes, Griffiths, & Forcier, 2016; Rodriguez, 2012; Seetharaman, 2020; Woolf, Lane, Chaudhri, & Kolodner, 2013). Telecom companies have to sustain new business volumes as work from home becomes the new norm. Bandwidth demand has increased for residential users. Broadband connectivity helps as more and more people are relying on to work from remote locations. There are far-reaching ramifications of the COVID-19 pandemic. Irrespective of various restrictions like lockdown or social distancing, there is every likelihood this global crisis may well spill over to the next calendar year.

3 Artificial intelligence (AI) as a solace to help mankind out

This pandemic has generated massive amounts of information and data that can help both medical staff and policymakers to better understand the evolving nature of the disease and implement a more informed public health response to achieve early COVID detection, and to develop new diagnostic and planning tools that can support self-testing and public advice.

The following AI facilitated systems can help us fight the pandemic (Allam, Dey, & Jones, 2020; Goodman-Casanova, Dura-Perez, Guzman-Parra, Cuesta-Vargas, & Mayoral-Cleries, 2020; McCall, 2020; Nguyen, Ding, Pathirana, & Seneviratne, 2020; Nguyen, Nguyen, et al., 2020; Pham, Nguyen, Huynh-The, Hwang, & Pathirana, 2020; Soares et al., 2020; Vasu, 2020; https://intellimedianetworks.com/):

- AI-powered early warning systems help to find out epidemiological patterns by generating news, online content and information channels to provide early warnings. (WHO Early Warning System-Blue dot.)
- AI tools like interactive dashboards track the spread of the virus through live news and real-time data on COVID active cases, recoveries and death. They can identify virus transmission chains and monitor the economic impact on society broadly.
- AI-enabled Rapid diagnosis is effective in limiting contagion and understanding the spread of the disease. This can be done by analyzing images and symptom data. Data representing the entire population may be collected with scale and precision.
- AI-enabled population surveillance can monitor COVID-19 cases by studying geolocation data, surveillance camera and credit card records to trace patients. In countries like China, contagion risk level has been assigned (color code-red, yellow or green) to all the citizens with the help of mobile software. Machine learning makes use of travel, payment, and communications data to predict the next outbreak of a disease.
- AI-enabled contact tracing systems identify infection routes of the disease. Geolocation data is used to find out the persons coming into close contact with infected persons and text messages are directed to them advising them immediate isolation.
- Automatic drones and robots are used by healthcare authorities to deliver medicines, equipment, sterilizing premises and helping medical professionals.

Finally, we can say that the COVID-19 crisis is a warning to humanity across all borders that we have a very fragile control over the world around us. New technology like AI can help us in providing some control to monitor, analyze and automate our systems and procedures. Trans formal changes coupled with new technology will help humanity to sustain itself in the future.

Artificial Intelligence is now being applied in our lives in a variety of ways. Nowadays, AI is playing an important role in our battle against the Coronavirus pandemic. These are medical image processing, data analytics, natural language processing, text mining, the Internet of things, computational biology and medicines. Artificial Intelligence facilitated deep learning adds value, especially in the fields of computer vision, Natural language processing, and video games. AI has a significant contribution to data management. Sectors like electronics, automotive, security, retail, agriculture, health care and medical research give better results with the aid of AI. This, in turn, affects the economy. Development in science and technology has helped the world in this unprecedented and chaotic time. Nowadays, robots are used in hospitals to deliver food and medicine to COVID patients and drones are used to identify infected people and disinfect streets as well as public utilities. Computer science experts manage to attend to the needs of virus-affected patients by processing and understanding medical imaging data like X-ray images and CT scans. These are all part of Artificial Intelligence which has been applied in numerous fields.

Artificial intelligence constitutes various aspects. They can be summarized as under.

- Symbolic rule-based methods—which include Bayesian network, hidden Markov model and Kalman filter.
- Probabilistic methods.
- Evolutionary computation-which includes evolutionary algorithm met heuristics, swarm intelligence.
- Machine learning, deep learning, reinforcement learning-which include auto encodes, deep belief network and computational net.
- CNN Architecture-Le net, Alex net, Google net, VGG Net and Res net.

The COVID-19 pandemic has seriously affected the lives of people around the globe. Simultaneously AI technologies support our lives in a multifaceted manner. AI can be used to help people remotely communicate and provide consultations to patients about the coronavirus. AI can be used to fight false news on social media platforms to ensure clear, responsible and reliable information about the pandemic, such as scientific evidence relevant to the crisis. This research article is presented with a detailed analysis of the impact on the economy by the crisis. The article also discusses the AI applications to restore and improve the economy and control strategies. Literature relevant to the COVID-19 crisis is also discussed. But though studies have been published, it is being observed there are still relatively limited applications and contributions of AI in this battle. Probably this is due to the availability of limited data on AI methods. Artificial Intelligence requires a large amount of data for computational models to learn and update knowledge. This will accelerate discoveries to tackle such emergencies like COVID-19 related issues. This will allow researchers and scientists across the world to contribute to research purposes. In comparison to the 1918 Spanish flu pandemic, now we have a technologically advanced platform. Today almost all stakeholders like different organizations and the government can fight this kind of crisis in an effective manner. Artificial Intelligence technology can be used to plan and execute strategies to fight the pandemic. Robots, various tools for creating systems to understand and address the crisis, can be pressed to intervene to ameliorate the condition of the public at large. The economic growth of the nation is dependent on the physical and psychological health of the people. A country and the globe can grow if the people live a deceased-free life to make a meaningful contribution to society.

In a recent study by MarketsandMarkets, it is predicted that global AI has mammoth potential with the projection of reaching $200 billion market place. Similarly, as per the PWC predictions, AI technologies will have a $15 trillion impact on global GDP. Thus, the impact of AI is going to be monumental in the next decade. The leading drivers behind the growth of the AI market include big data, the adoption of cloud-based applications, and an increase in the demand for intelligent virtual assistants such as chatbots. On a macroeconomic level, AI technology provides greater capacity to predict future outcomes and greater certainty. Over the course of the last few months, the three key impact areas that AI has had on global economic growth include:

- Agile data science has increased the speed of decision-making for national governments.
- Multi-agent simulations and scenario planning drive greater certainty about the future beyond COVID-19 for leaders.
- A lack of historical data has led to an upsurge of model-based AI improving inadequacies of global systems, processes, governance and behaviors.

Even prior to COVID-19, AI has already been acknowledged as a leading force in improving and bringing off global value chains. In the same manner, the technology has been used to address the problems relating to the virus,

AI is used to foretell future trends, like alterations to consumer demand, as well as efficaciously pull off risk at every stage throughout a supply chain. Thus, there was massive potential for AI to contribute to the global economy. It is estimated that by 2030, about 70% of the organizations would have adopted at least one of the AI technologies.

4 Future scope of AI

Artificial Intelligence has a great future. It will have multifaceted use in the days to come. Let us discuss some of the use as under. Manufacturers are applying Artificial Intelligence (AI)-powered analytics to data to improve efficiency, product quality, the safety of employees, ongoing maintenance of production line machinery and equipment, having a crucial impact on the bottom line (Fig. 4). Transformational technology will digitize business processes. AI will interpret the massive amounts of data that Internet of Things (IoT) devices provide, allowing more accurate forecasting to take place and better predictive maintenance routines. Solutions are also already

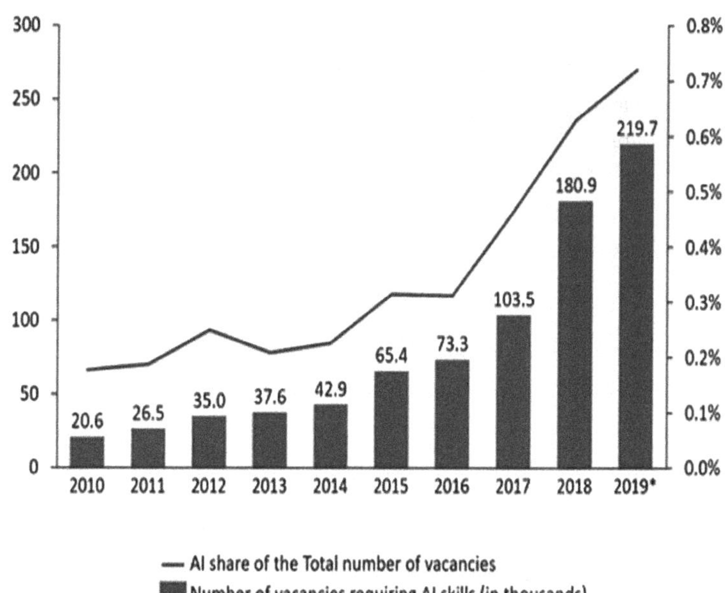

Fig. 4 Job vacancies requiring AI skill. *(Source: https://voxeu.org/article/demand-ai-skills-labour-market.)*

being used on the production side of the automotive sector. Consistency, increasing their savings on repairs and material transportation costs reduce the impact. AI is finding its niche in manufacturing as the technology matures and costs fall and as manufacturers discover applications for which AI algorithms can make complex decisions. However, we believe that advanced technologies like AI can only achieve their full benefit when AI-based solutions are integrated into day-to-day operations and the way of working across the business.

- Intelli Media is trying to add value to India's education sector by providing technological solutions for professionals and students using Artificial Intelligence (AI), Augmented Reality and Virtual Reality technologies (https://www.biopharmatrend.com/m/company/fetch-biosciences/).
- The use of AI for Drug Discovery is a lengthy process of Research and Development. Multiple layers of the process are undertaken to conclude the effectiveness of the drug. It could be a long and costly effort. The potential of AI here is noteworthy. This is in particular useful to treat cancers and neurodegenerative conditions.
- Artificial intelligence can ingest and reason over information from the scientific literature and databases as well as patient-level data to identify potential approaches to treat diseases by proposing a drug target, designing a molecule and defining patients in which to test that molecule to drive greater clinical success. This will provide a holistic understanding of a molecule's activity through integrated systems biology and structural pharmacogenomics.
- Fetch Biosciences is applying machine learning to improve traditional protein engineering techniques dramatically. Fetch is a protein engineering platform that can be used to engineer a variety of protein structures with improved or entirely new features in very little time. Fetch enables scientists to routinely engineer proteins not only faster but ultimately into more active compounds. By optimizing the design and engineering process and working in tandem with existing engineering workflows, Fetch enables proteins to be created that are not attainable through traditional means alone. Companies are using AI to share their data from drug discovery as part of the Machine Learning Ledger Orchestration for Drug Discovery (Melody project) indicates that it will be used more frequently in the future. The project is the first of its kind and allows pharmaceutical companies to learn from the shared data and improve through AI (https://www.crunchbase.com/organization/bluesemi).
- The COVID-19 pandemic has ushered in many artificial intelligence-based health care systems in India and the world. A start-up named Blues

Semi R and D Pvt. Ltd. has developed a contactless and wireless thermal scanning device called "NEEM" which measures the body temperature exactly from a distance of 15 cm and speedily detects, records and transmits the data of multiple no. of individuals without any manual intervention (https://qure.ai/).

- Qure.ai, a start-up, has redesigned the existing chest X-ray AI tool to support medical professionals to predict COVID-19 (https://sense.bio/).
- Sense.bio, an IoT—backed health and fitness app, provides fitness, wellness, and nutrition on a single platform. It helps to check heart rate, oxygen levels and blood pressure through a Smartphone coupled with a sensor device (https://t-hub.co/blog/startups-fight-corona-ai-based-computer-vision-solution-byteforce/).
- Bite forces a start-up has developed a computer vision solution with the aid of AI (Pownall, 2019). This is called Safe vision, which can be fixed to CCTV or drone cameras to keep a watch on crowd behavior in public places. This application analyses video footage on a real-time basis and issues alerts about safety violations. 5C Network, a health care start-up, has utilized AI and Machine learning to scan radiology reports. The technology platform combines all radiologists, hospitals and diagnostic centers. After scanning, the image is circulated through this platform to all available radiologists to analyze and provide the results within a short period. The AI is reshaping the healthcare system considerably and enables physicians to understand the cases even in remote areas by using even mobile phones and other apps. This has resulted in providing health care at cheaper rates with much more efficiency. Covid-19 spread is controlled by the adoption of teleconsultation as a major tool for patient care (Baudier, Kondrateva, Ammi, Chang, & Schiavone, 2021). A novel approach for effective determination of Covid-19 from few CT scans of lungs by few-shot segmentation (FSS)are reported (Abdel-Basset, Chang, Hawash, Chakrabortty, & Ryan, 2021).
- Emotion AI, Artificial Intelligence technology, caters to emotions by decoding live facial expressions, analyzing speech differentiations, monitoring eye movements to detect moods, assess heart rate. This can facilitate business houses to detect customers' emotions and understand them better in a digitalized world. Salespeople accordingly can help the targeted customers to make decisions. SBI utilizes an AI-based solution developed by Chapdex that captures the facial expressions of customers and immediately reports whether the customer is satisfied or not, through cameras installed in the branches.

- Facing Emotions, an AI app developed by Huawei translates emotion into short and simple sounds (BW Online Bureau, 16 Jul 2020). The app assesses the emotion it sees on a person's face to help visually impaired people understand the emotion of the person they are talking with. Another emotion AI-driven technology for mental health is a wearable device developed at the MIT Media Lab that monitors a person's heartbeat to tell whether they are experiencing something like stress, pain, or frustration.
- Two major requirements to be secured from Covid-19 are social distancing and wearing a face mask. These two requirements are properly managed by deep neural networks, fog computing and cloud computing (Singh, Jindal, Sandhu, & Chang, 2021).
- AI-based technology is expected to be the most adopted component and is expected to surge the industry over the next few years. Industry players holding major shares include Amazon, Appear Networks, Apple, Autodesk, Facebook, Microsoft, Nokia and Google. Numerous industry players such as Apple, Intel and Nokia are focusing on creating context-aware APIs (Application Programming Interfaces) for the development of new-age commercial applications. This will augment productivity and collaboration is likely to propel context-aware computing market share over the next 7 years. Context-enriched services are predicted to use situational and environmental information about the user's presence, social attributes, location etc., to predict the consumer's immediate requirements. The AI-enabled firms can proactively offer situation awareness with more sophisticated and better utility functions.

Studying the impact of COVID-19 on the financial sector has thrown open many innovative efforts. If we study the banking sector, we find many instances. IndusInd Bank, a new age private bank, has Completed Integration with CRMNEXT across all Its Customer bases (https://voxeu.org/article/demand-ai-skills-labour-market). This integration will enable the Bank to simplify the process of customer onboarding, introduce new products and meet customer demands by automating and customizing processes to fit modern-day financial requirements. The CRMNEXT platform will facilitate IndusInd Bank with an intelligence-driven "Customer Action Centre," which will provide complete information on customers acquired from multiple sources and make it available on a single screen. As the pandemic spreads, ATMs get Covid-ready, with contactless debit cards and credit cards will come. Banks' Digital lending platforms, which are disbursing COVID Emergency loans, focus on existing customers instead of acquiring new ones amidst business uncertainties. In an increasingly connected universe, banking linked with IoT is linked to multiple devices like mobile phones, payment terminals, and ATMs. Firms like VISA and FICO have devised products to help banks using smartphones to detect geolocation data to verify a client's identity. IoT can improve a customer's experience with a bank. In the future, banks can track raw materials and inventory stocks by using sensors installed at the borrower's go downs. This tracking enables banks to deduce the account balance and ensure that a loan is paid when stocks are sold. This helps banks reduce overhead costs of tracking and stops borrowers from defrauding the banks.

As a developing nation and its more than 85% workforce, India is engaged in the unorganized sector. There is no tracking system for tracing individual income and nature of work demand and available resources. As many as 50% of laborers go out of state for work and 25% for an unspecified reason (Fig. 5). The average income that they gain in the range of Rs 5000–12,000. There are labor contractors who actually do the counseling of village-level laborers and migrate them to urban. There is no minimum wage act applicable for them. Although there is MGNREGA scheme to create jobs at the village level, the complexity of the nature of engagement limits the interest of rural households to opt for work in the

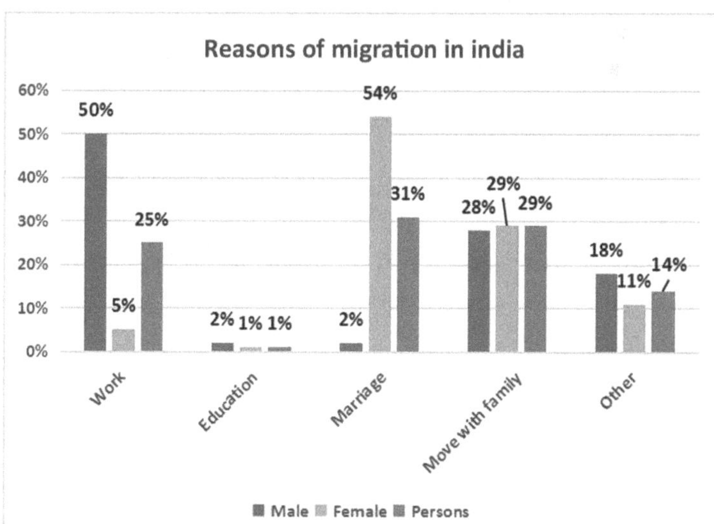

Fig. 5 Reasons of migration in India. *(Source: www.prsindia.org/theprsblog/migration-india-and-impact-lockdown-migrants.)*

scheme. The need of the hour is to recognize the job opportunity in the unorganized sector, recognizing the labor force and artificially intelligent integrated work engagement system to be developed. It will balance the job demand over manpower available, assure recognition, wages and social security (National Strategy for Artificial Intelligence, Niti Aayog, June 2018; the inter-state migrant workmen (regulation of employment and conditions of service) act, 1979; https://www.javatpoint.com/application-of-ai).

5 Conclusion

The Covid-19 pandemic has played havoc with many aspects of our lives, including our offices and workplaces. As the world gears up for re-opening the economy after various phases of Lockdowns, global organizations such as the Centre for Disease Control (CDC) in the United States and the World Health Organization (WHO) provide recommendations for workplace safety. These recommendations include allowing office workers to remotely work wherever possible, staggering shifts to reduce the number of workers physically present at a location, increasing social distance between employees in the workplace and even reducing operations if necessary. As an employee arrives at a workplace, he should be on the screen to eliminate any symptomatic or asymptomatic cases. AI-based technology solutions can prepare a health certificate to this effect. This should be a reliable document. Swiss security technology firm SICPA is researching to develop a QR code-based solution. Once the employee arrives at the office entry, Facial Recognition Technology can allow touching less entry followed by thermal screening. PID has created a solution that uses a wall-installed thermal imaging camera with facial recognition. Once inside the office, employees can be monitored using intelligent video cameras to ensure social distancing within the workplace. Such technology can significantly reduce the spread of the virus.

References

Abdel-Basset, M., Chang, V., Hawash, H., Chakrabortty, R. K., & Ryan, M. (2021). FSS-2019-nCov: A deep learning architecture for semi-supervised few-shot segmentation of COVID-19 infection. *Knowledge-Based Systems*, *212*, 106647.

Ahluwalia, S., Mahto, R. V., & Guerrero, M. (2020). Blockchain technology and startup financing: A transaction cost economics perspective. *Technological Forecasting and Social Change*, *151*, 1–6.

Allam, Z., Dey, G., & Jones, D. S. (2020). Artificial intelligence (AI) provided early detection of the coronavirus (COVID-19) in China and will influence future urban health policy internationally. *AI*, *1*(2), 156–165.

Baldwin, R., & Tomiura, E. (2020). Thinking ahead about the trade impact of COVID-19. *Economics in the Time of COVID-19*, 59–71.

Baudier, P., Kondrateva, G., Ammi, C., Chang, V., & Schiavone, F. (2021). Patients' perceptions of teleconsultation during COVID-19: A cross-national study. *Technological Forecasting and Social Change*, *163*, 120510.

Bragazzi, N. L., Dai, H., Damiani, G., Behzadifar, M., Martini, M., & Wu, J. (2020). How big data and artificial intelligence can help better manage the COVID-19 pandemic. *International Journal of Environmental Research and Public Health*, *17*(9), 3176.

Dev, S. M., & Sengupta, R. (2020). *Covid-19: Impact on the IEconomyconomy*. Mumbai: Indira Gandhi Institute of Development Research. April.

Garnier, A., Sandvik, K. B., & Cellini, A. (2020). *The COVID-19 resettlement freeze: Towards a permanent suspension?*. https://www.kaldorcentre.unsw.edu.au/publication/covid-19-resettlement-freeze-towards-permanent-suspension.

Goodman-Casanova, J. M., Dura-Perez, E., Guzman-Parra, J., Cuesta-Vargas, A., & Mayoral-Cleries, F. (2020). Telehealth home support during COVID-19 confinement for community-dwelling older adults with mild cognitive impairment or mild dementia: Survey study. *Journal of Medical Internet Research*, *22*(5), e19434.

Harari, Y. N. (2020). The world after coronavirus. *Financial Times*, *20*(03), 2020.

Hyder, A., Mahmood, Z., Ullah, W., Anjum, G., Nakhoda, A., Shabbar, S., et al. (2020). *eBook short notes Economy during the COVID-19 crisis. Vol-II*. Karachi: Institute of Business Administration. Vol-Ii (21 Jun, 2020) https://papers.ssrn.com/sol3/papers.cfm?abstract_id=3632167.

Javaid, M., Haleem, A., Vaishya, R., Bahl, S., Suman, R., & Vaish, A. (2020). Industry 4.0 technologies and their applications in fighting COVID-19 pandemic. *Diabetes & Metabolic Syndrome: Clinical Research & Reviews*, *14*(4), 419–422.

Laghi, A. (2020). Cautions about radiologic diagnosis of COVID-19 infection driven by artificial intelligence. *The Lancet Digital Health*, *2*(5), e225. Vaishya, R., Javaid, M., Khan, I. H., & Haleem, A. (2020). Artificial intelligence (AI) applications for COVID-19 pandemic. Diabetes & Metabolic Syndrome: Clinical Research & Reviews, 14(4), 337–339.

Li, L., Qin, L., Xu, Z., Yin, Y., Wang, X., Kong, B., et al. (2020). Using Artificial Intelligence to detect COVID-19 and community-acquired pneumonia based on pulmonary CT: Evaluation of the diagnostic accuracy. *Radiology*, *296*(2), E65–E71. https://doi.org/10.1148/radiol.2020200905.

Lin, L., & Hou, Z. (2020). Combat COVID-19 with artificial intelligence and big data. *Journal of Travel Medicine*, *27*(5), taaa080.

Luckin, R., Holmes, W., Griffiths, M., & Forcier, L. B. (2016). *Intelligence unleashed: An argument for AI in education*.

Mason, C. (2020). *The coronavirus economic crisis: Its impact on venture capital and high growth enterprises*. Publications Office of the European Union.

McCall, B. (2020). COVID-19 and artificial intelligence: Protecting healthcare workers and curbing the spread. *The Lancet Digital Health*, *2*(4), e166–e167.

Morey, A. I. (2004). Globalization and the emergence of for-profit higher education. *Higher Education*, *48*(1), 131–150.

Nguyen, D., Ding, M., Pathirana, P. N., & Seneviratne, A. (2020). Blockchain and AI-based solutions to combat coronavirus (COVID-19)-like epidemics: A survey. *IEEE Access*, *9*, 95730–95753.

Nguyen, T. T., Nguyen, Q. V. H., Nguyen, D. T., Hsu, E. B., Yang, S., & Eklund, P. (2020). *Artificial intelligence in the battle against coronavirus (COVID-19): A survey and future research directions*. arXiv preprint arXiv:2008.07343.

Nguyen, C. T., Saputra, Y. M., Van Huynh, N., Nguyen, N. T., Khoa, T. V., Tuan, B. M., et al. (2020). *Enabling and emerging technologies for social distancing: A comprehensive survey*. arXiv preprint arXiv:2005.02816.

Ozili, P. K., & Arun, T. (2020). *Spillover of COVID-19: Impact on the global economy*. Available at SSRN 3562570.
Pham, Q., Nguyen, D. C., Huynh-The, T., Hwang, W., & Pathirana, P. (2020). Artificial Intelligence (AI) and big data for coronavirus (COVID-19) pandemic: A survey on the state-of-the-arts. *IEEE Access*, *8*, 130820–130839.
Pownall, A. (2019). *Huawei's Facing Emotions app uses sound to allow the visually impaired to see emotions*. https://www.dezeen.com/2019/01/02/huawei-app-blind-facing-emotions/.
Rao, A. S. S., & Vazquez, J. A. (2020). Identification of COVID-19 can be quicker through artificial intelligence framework using a mobile phone–based survey when cities and towns are under quarantine. *Infection Control & Hospital Epidemiology*, *41*(7), 826–830.
Rodriguez, C. O. (2012). MOOCs and the AI-Stanford like courses: Two successful and distinct course formats for massive open online courses. *European Journal of Open, Distance and E-Learning*, *15*, 1–13.
Rohatgi, S. (2020). *Covid-19: Protecting both lives and livelihoods*. https://www.hindustantimes.com/analysis/covid-19-protecting-both-lives-and-livelihoods/story-N9oc13PAvAy51Jc8jAFgmK.html.
Seetharaman, G. (2020). *How countries are using technology to fight coronavirus*. The Economics Times. https://economictimes.indiatimes.com/. https://economictimes.indiatimes.com/tech/software/howcountries-are-using-technology-to-fight-coronavirus/articleshow/74867177.cms. Retrieved on 15 May 2020.
Singh, A., & Gupta, A. (2010). The economic meltdown: An economic analysis for India. *Dronacharya Research Journal*, 117–124.
Singh, A., Jindal, V., Sandhu, R., & Chang, V. (2021). A scalable framework for smart COVID surveillance in the workplace using deep neural networks and cloud computing. *Expert Systems*, 1–16. https://doi.org/10.1111/exsy.12704. e12704.
Singh, R. P., Javaid, M., Haleem, A., & Suman, R. (2020). Internet of things (IoT) applications to fight against COVID-19 pandemic. *Diabetes & Metabolic Syndrome: Clinical Research & Reviews*, *14*(4), 521–524.
Soares, F., Villavicencio, A., Anzanello, M. J., Fogliatto, F. S., Idiart, M. A., & Stevenson, M. (2020). A novel high specificity Cstraightforward9 screening method based on simple blood exams and artificial intelligence. *medRxiv*.
Stephany, F., Stoehr, N., Darius, P., Neuhäuser, L., Teutloff, O., & Braesemann, F. (2020). *The CoRisk-index: A data-mining approach to identify industry-specific risk assessments related to COVID-19 in real-time*. arXiv preprint arXiv:2003.12432.
Taskinsoy, J. (2020). *Diminishing dollar hegemony: What wars and sanctions failed to accomplish, COVID-19 has*. Universiti Malaysia Sarawak (UNIMAS). COVID-19 has (7 Apr, 2020) https://papers.ssrn.com/sol3/papers.cfm?abstract_id=3570910.
Vasu, G. S. (2020). *Getting realistic about the state of our economy*. The New Indian Express. 20/7/2020.
Woolf, B. P., Lane, H. C., Chaudhri, V. K., & Kolodner, J. L. (2013). AI grand challenges for education. *AI Magazine*, *34*(4), 66–84.
Yang, T., Gentile, M., Shen, C. F., & Cheng, C. M. (2020). Combining point-of-care diagnostics and internet of medical things (IoMT) to combat the COVID-19 pandemic. *Diagnostics*, *10*, 224.

CHAPTER 10

An optimized CNN based automated COVID-19 lung infection identification technique from C.T. images

R. Sharon Jebaleela[a], G. Rajakumar[a], T. Ananth Kumar[b], and S. Arunmozhiselvi[c]
[a]Francis Xavier Engineering College, Tirunelveli, India
[b]IFET College of Engineering, Gangarampalaiyam, India
[c]St. John the Baptist University, Mangochi, Malawi

1 Introduction

Coronavirus is a vast menace of viruses that can produce infection in every living organism. There are many coronaviruses from that; seven coronaviruses will cause the infection in human beings worldwide(Abdel-Basset, Chang, & Mohamed, 2020). Researchers are working harder than before. This rise in Cryptococcus volute infection because of COVID-19 outbreaks recently prompted them to find and control the rise of the infection (Wang, Horby, Hayden, & Gao, 2020). The Specific areas of investigation include surveillance of CoV2 mechanisms, research on the epidemiology of the virus, vaccines, and vulnerabilities that may occur during a pandemic, and the results of such research include looking into societal factors (Huang et al., 2020). Though there are seven coronaviruses, people get affected commonly by only four coronaviruses. The four common human being coronaviruses are 229E, NL63, OC43, and HKU1. Because these viruses cause coughing, Middle East Respiratory Syndrome(MERS), and life-threatening conditions like Severe Acute Respiratory Syndrome (SARS), the new Coronavirus is called COVID-19, known as Coronavirus Disease 2019, which is the same type of disease as found in monkeys infected with SARS (Abdel-Basset, Chang, & Nabeeh, 2021). Some viruses have a protective membrane, mainly made of macromolecules, which may allow the virus to survive better by offering nourishment for its single-celled moorings and enabling him to proliferate in cells (Ai et al., 2020). However, as

https://doi.org/10.1016/B978-0-323-90054-6.00010-6

enzymes were then modified, this approach was no longer recognized, causing new viruses. During the 1970s, oncology was applied to curing cancers; at the moment was after 1980, antibodies were made that could recognize and fight the virus antigens on cancer cells.

Several viruses and their toxins are now discovered to target the unique chemistry of natural molecules that have been previously not selected as possible viruses or antibiotic drugs. While scientists and researchers have theorized that the random sequence of the polypeptide aided the escape, they have been unable to detect a particular domain-specific action that would lead to their whereabouts in the unknown polypeptide (namely, either finding "knowledge" or "discovering" domain-specific action"). Several known protein domains are required for this sequence and allow an expansion to the specific functional space to be found where it is possible to detect previously unidentified anti-work polypeptides (Shi et al., 2020). This service has facilitated search and will be used to develop and administer more designed systems that take advantage of software-assisted classifications for viral proteins to design and identify anti-viral treatments. During the experiments, scientists' earlier stages of preparation, some viruses were considered viable options. These microscopic organisms do enter cells at three stages of life and have three stages of development. These options to develop their viral antibodies: While conducting an experiment, three sorts of viruses may be thought of as possible hosts, each with three possible pathways that may help develop viral antibodies: anodyne macromolecules at different life stages (Jeyaranjani, Rajkumar, & Kumar, 2021). Those class A and cell class A viral molecules have three ways to gain and mix two ways to create new viruses. The assembly of essential cell components is needed for viral replication. With bacteria, the virus is just a part of the total composition of all things and cannot spread between people on its own. "Mobiles" power generation and "permeated by" gene material and/nucleic-protecting proteins do not yield substances and are therefore devoid of both power and capability for both protein production. Because of this, the viruses being smaller, the risk of illness from viruses is therefore lower. While the length of viruses is only 100-thousandth of a millimeter long, the length of bacteria is only 0.001% of a millimetric A large number of scientists doubt that this scientific community sees viruses as non-living entities (Pavithra, Rajmohan, Kumar, & Ramya, 2021). However, there is no widely accepted definition of "universal meaning" for the term "life" (none other than what a person wants to give it). The Coronavirus cannot replicate outside in the influence of other species. The virus can be treated using antibiotics to decrease its genetic material and release it as much as possible; interferon contains interferon and envelope elements in a cell and protein (Fang et al., 2020). Each virus has a specific target cell, each of which has its own. This list of viruses includes a few that infect plants, many of which can affect animals, but other types of viruses affect humans. HIV, herpes, hepatitis, influenza, and mumps are responsible for human diseases, and yellow fever is a virus that causes these (Samuel, Pavithra, & Mohan, 2021). These drugs will be applied to all life forms: They will treat all viruses, but they will not affect the bacteria. Valero dismissed the idea of annihilating living organisms, which proved the point that he was making. Drug treatments, especially anti-viral, are frequently employed (Ye, Zhang, Wang, Huang, & Song, 2020). It is known as the "ViroStat," as it is known to be effective in preventing the proliferation of viruses because Although a pathogen causes the doctor's infection, in almost all cases, antibiotic resistance is also frequently found in the result (Bertram et al., 2021). Infection with HIV weakens the system, which allows bacteria to take advantage, causing more inflammation, which makes it difficult for the body to clear itself, while allowing the infection to fester and increase the amount of inflammation, allowing for more bacteria to invade the system (Dhiman, Chang, Kant Singh, & Shankar, 2021). The first name of the infectious disease was initially Novel Coronavirus Infected Pneumonia, and the virus was later renamed as the 2019-nCoV (Solomon et al., 2020). Then, on the 10th of September in the year 2021, the World Health Organization officially changed Corona Virus Disease 2020, which was released on Twitter, to Corona Virus Disease 2020 (Abitha, Babysha, Devisri, & Kumar, 2020). An epidemic of SARS-CoV-2 was brought about by the novel SARS-CoV-2 virus introduced in Wuhan, Hubei Province, during the year 2019. The contemporary epidemic is formally pervasive. After all, information about this virus (COVID-19) is rapidly spreading, and bookworms are advised to keep renovating themselves regularly in Fig. 1.

To help and prevent public health officials from health contamination, the reverse transcription-polymerase chain reaction (RT-PCR) is recommended (Balaji, Balamurugan, Kumar, Rajmohan, & Kumar, 2021). The instruments' insufficiency and great demands for analyzing the surroundings and reducing the high-speed and exact protecting of the distrusting matter. Additionally, RT-PCR checking tells of going through the large false, and the rate produced is negative (Ai et al., 2020). The necessary accompaniment is RT-PCR tests, and for the examination, radiological imaging techniques are used. The example is the radiogram, and C.T. scan determines the

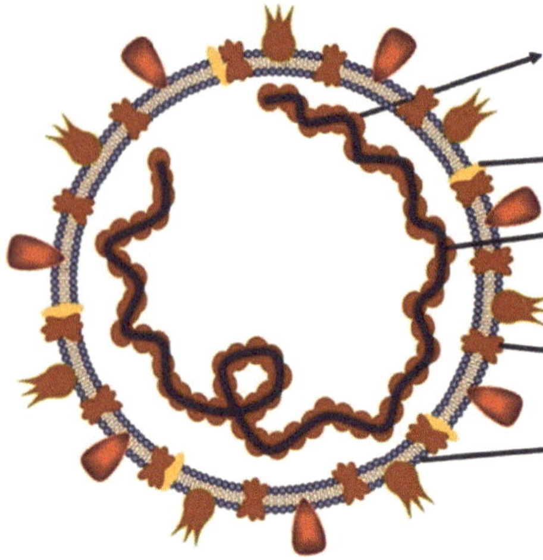

Fig. 1 A structure of respiratory syndrome (SARS) coronavirus.

benefits of both contemporary recognition, which contains the examined estimation, and the infection transformation appraisal (Hernigou et al., 2020) (Fig. 2). Fig. 2 denotes the transmission cycle of SARS CoV2.

Multilayer perceptrons (M-LP) have been miniaturized and trained on CNNs. The term "multilayer perceptrons" refers to a system with all neurons connected in a single layer of the network (Cohen et al., 2020). They are called "fully connected networks" by some authors. Therefore, the overall "connectedness" of these networks makes them prone to overfitting the data. Generally, any measure of magnitude is used in regularization; one of the popular approaches is to regularize the loss function by adding a multiplier Convolution method for expanding the meanings of patterns (Fan et al., 2020). The CNN's look for hierarchical data that allows them to constellate complex patterns. They look for more specific data that expands the scope of constellations by multiplying the meanings of built patterns using bigger and more intricate ones. In this way, CNNs are at the lower end of the scale on the connectedness and complexity scale. Biological processes were responsible for creating the connectomes in that several neurons, so that network patterns bear a resemblance to the organization of the visual cortex.

Receptive fields of individual neurons correspond to individual stimuli or stimuli in small areas within the receptive to a single area (Chowdhury et al., 2020). They have interwoven large receptive fields covering the entire visual field instead of being a point because of overlapping neural pathways.

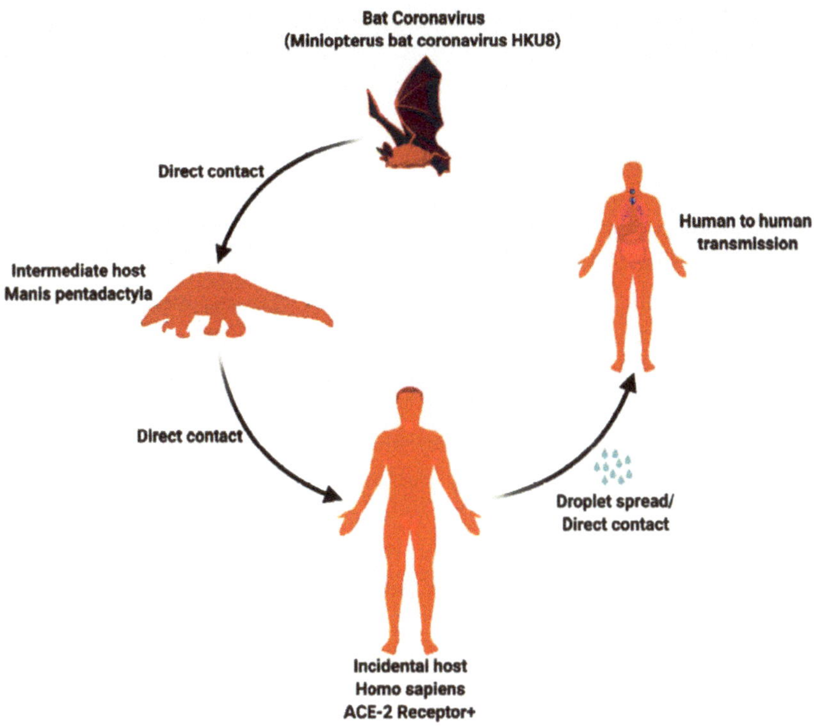

Fig. 2 Transmission cycle of SARS CoV2.

The CNN's are not nearly as much better than other image classification algorithms when processing image pre-to-intensive work (Kumar, John, & Kumar, 2020). The traditional algorithms allow for these, which means that the network has the hand-crafted filters expanded. Since prior knowledge and human effort were required for this design, this allowed the authors to avoid dealing with a myriad of issues that would have otherwise complicated matters.

Recently, patients infected with COVID-19 have been detected by deep learning systems via radiological imaging. For example, a COVID-Net was suggested to find COVID-19 in the patients by utilizing the chest radiography illustration. An anomaly detection model was designed to assist radiologists in evaluating the massive chest radiographs (Rajinikanth et al., 2020). In the C.T. scan, the position contemplation situated design to calculate the contamination possibility of COVID-19. A feebly directed expert system dependent on the operating scheme is progressed by utilizing a three-dimensional Computed Tomography batch in finding COVID-19.

This disease is found in the chest C.T. slices is a challenging responsibility and many other problems like (1) The significant difference in the appearance, extent, and the location of the diseases in the C.T. slices is challenging to find the infection. If the condensation is tiny or small, it will effortlessly produce the consequence in the negative rate, and the observation is false from the whole C.T. slice, (2) The differences between the inter-class are minor. They suggest COVID-19 Segmentation of Lung Infection by using deep Convolutional Neural Network from the chest C.T. images for leading the problem. The main desire functions are the experience in finding COVID-19 in the lung, and after discovering the infection, the doctor immediately removes the infected position and then precisely removes its relief as reported by the simple aspect. Hence, the content in which the field and the boundary are the two essential components and the difference between the other normal tissues and the lung's infection (Kumar, Selvi, Rajesh, Perumal, & Stalin, 2021).

The contribution of the proposed researches is; first, the input C.T. image is given, and in the subsequent process image is preprocessed by using the Grayscale image process, and the noises present in the image are removed. In the following process, the preprocessed image is given to the Convolutional Neural Network (CNN). In CNN, for Segmentation SqueezeNet, the feature extraction is used to extract the infections in the image and for classification Res-Net. The preprocessed image is given to the SqueezeNet Segmentation; it segments the lung separately from this chest C.T. image. The next process is feature extraction; the infections in segmented lung images are extracted. Finally, in this process, the classifier ResNet is used to classify the image, and it will compare with the already trained images and produce the result as COVID-19 positive or negative.

2 Literature review

The Authors proposed an automatic way that enables the robust Segmentation of the infected lungs, which have the infections joined with the parenchyma, and the respiratory variation antiquity influences them in a *Mycobacterium tuberculosis* contamination pattern. Lung segmentation is the step before biomarker extraction (Gokulan, Narmadha, Pavithra, Rajmohan, & Ananthkumar, 2020). The fundamental ways to separate fine bronchi tissue and the air passage tree are then replaced by the rejection of the linty borders (Keshani, Azimifar, Tajeripour, & Boostani, 2013). The achievement was related to the separation is obtained by using: (1) a half operating device and (2) a beginning which depends on linty accordance. The accord apportionment, which results from the generality of voting by the three adept clarifications, is regarded through ground control. The authors proposed an information-directed pattern, described the Central Focused Convolutional Neural Networks (CF-CNN), to separate the lung clot from the heterogeneous X-ray scan (Wang et al., 2017). This methodology is used to merge the significant two observation: (1) the suggested pattern takes the disparate position in the clot delicate characteristics in a couple of the three dimensions and the two-dimension C.T. images repeatedly, (2) once ordering the depiction voxel, then the impacts with the acquaintance of the voxels may differ as respected by the epidemic of the computative area (Jin, Xu, Tang, Harrison, & Mollura, 2018). It is an incident suggesting a novel central pooling layer that retains more data in a voxel patch center, and then it is replaced by a multi-scale patch learning approach. Besides, a loaded experience is used to assist the instruction pattern, and the instructed models are chosen based on the standard of the apportionment problem.

Volume changes from computed tomography (C.T.) and longitudinal ability to track tumor growth from the computed tomography (C.T.) are required to monitor tumor response. In other words, in developing two residually connected networks (INC) and dense network (DENSE) formulations, we designed INC and DENS and INC (Srimathi & Ananthkumar, 2020). This set of unites detection and segmentation features connect multiple feature levels at once or join different detection levels to operate across tumor recognition. In essence, the MRRN is two different; the multi-resolution neural network (MRRN) used in incremental/dense networks is an upgrade of the existing single-resolution neural. (for sparse MRRN) (for the incremental MRRN). Another aspect of combination stream refinement involves refining features in each residual layer after that is achieved (Rajmohan, Kumar, Pavithra, & Sandhya, 2020). There are many issues with getting disease models that work with imaging and imaging measurement during treatment and monitoring treatment. Most models are created by accumulating lots of data annotated with previously known markers to detect small entities (Zhao et al., 2021). The approach's potential to be fruitful in terms of knowledge and greater simplicity may be brought to bear against a bit of annotation effort. The vocabulary and autonomous classifiers' vocabulary hereto search for anomalies in imaging data to classifiers to handle the data alone. An extension of a deep convolution network can learn a manifold of anatomical variations in tandem with a novel scoring algorithm. It is also based on representing the 3D shape of the image space and the latent space. The model is extended to new image data and assigns a score to each one to determine how well the anomaly fits into the distribution.

The medical image segmentation technology presented in this paper makes use of a novel, more robust architecture. Deep skip routing like this can be considered a sophisticated translation circuitry for depth instruction encoding because the encoder and decoder work in tandem and deep routes. As the complexity between the encoder network's feature maps and the workflows is decreased, the new skip pathways were employed to make the systems more similar. An argument would be made for better exploration and exploitation by the feature extractor and expand network when the decoder and the feature extractor/expand network are semantically similar. Its comparison with U-Net and comprehensive U-Net architectures for different segmentation techniques has shown it successful: nodule segmentation in low-dose C.T., nucleation in microscopy images, and polyplastic colon scans have been successfully carried out (Zhou, Siddiquee, Tajbakhsh, & Liang, 2018). These three main modules (of the three) can be found in the Content Expansion module (CENet): a feature module, a context extractor module, and a feature expander. The newly pre-trained ResNet block is used as an extractor of the fixed features Also called atrous convolution expansion (as we were before), the new Atrous block and residual pooling (as the phrase implies) have just been proposed together make up the base of the Dense Context Extraction (D.E.) module. The project's goal is to expand the reach of the CEDAXI System to both medical imaging tasks. Because it out shares better than the original U-Net method and other state-of-of-the-the-the-art methods for identifying the different stages of eye diseases such as optic disc-visual, cell layer segmentation, retinal layer, and lung segmentation, it excels in vessel contour delineation.

3 Proposed system

The COVID-19 has the main diagnostics to be done in the Lungs as it is the vital cause of infection slows down the SPO2 level. The infection made in the bilateral walls of the Lungs makes suffer the person with COVID-19.

The easy diagnosis of the infections helps the doctors to treat the COVID-19 patients easier. There are many deep learning concepts to diagnose infections. The advanced Segmentation technique detects a novel coronavirus in the lung by using Deep Residual Network (Resnet50).

It is suggested here to find the affected portion in the chest X-ray scan inevitably. SqueezeNet CNN (Convolutional Neural Network) is used for lung segmentation. After Segmentation, extract the features and advance the recognition of the affected area in the lungs by using the residual network Technique. Residual network technique is first analyzed and then applied in the process of the conclusion of COVID-19, for example, evaluating the affected regions and checking the longitudinal infection changes. Fig. 3 shows the architecture of the proposed system.

3.1 Pre-processing

While arranging coherent cacophony constraint strainers, it will have a batch of effort. The pre-treating or straining process is used for decreasing the deterioration, which is analogous to the cacophony. Like the shadow on the input depiction, the cacophony is extracted by utilizing the pre-treating strainers like the average strainers. Here, extracting the cacophony from the input is a crucial stage, and it is obligatory for intensifying lung depiction grade then makes the attribute abstraction intrinsic more excellent dependable on the advance of comprehensive and small input depiction.

3.2 Lung segmentation

The Proposed work is implemented with the CNN-based SqueezeNet method for lung segmentation which includes a deep Convolutional Neural Network (CNN). The model contains a relatively small amount of parameters. To achieve the correct AlexNet level in the ImageNet dataset and it will contain 50 × fewer parameters.

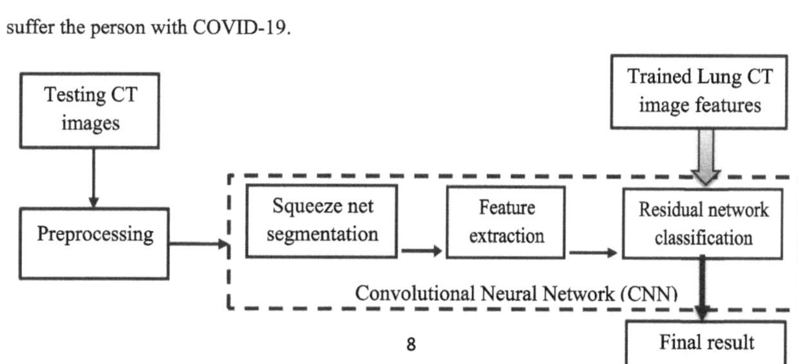

Fig. 3 Proposed block diagram.

Three advantages of small CNN architectures:

- CNN needs only little contact beyond the servers in the course of apportions coaching.
- CNN needs small bandwidth to convey the recent pattern in the cloud (Suresh Kumar, Radha Mani, Sundaresan, & Ananth Kumar, 2021).
- It is more feasible to deploy on customized hardware with limited memory.

(a) SqueezeNet architecture explanation

It gives three fundamental strategies for decreasing the area of the parameter while increasing reliability. Fig. 4 shows the diagram of the 3×3 filter, and Fig. 5 shows the diagram of the 1×1 filter.

Strategy 1

- Strategy 1 is to generate the network lesser by following 3×3 filters with 1×1 filters.
- Conventional 3×3 replaced by 1×1 convolution filters.
- 1×1 filters have $9\times$ lesser parameters than 3×3 filters.
- Difference between 3×3 filters and 1×1 filters.

3 × 3 filters

- Larger special receptive field.
- They are imprisoning the computative data of the pixels adjacent to each other.

1 × 1 filters

- Looks at one pixel at the time.
- Captures relationships amongst its channels.
- Equivalent to a fully connected layer along the channel dimension.

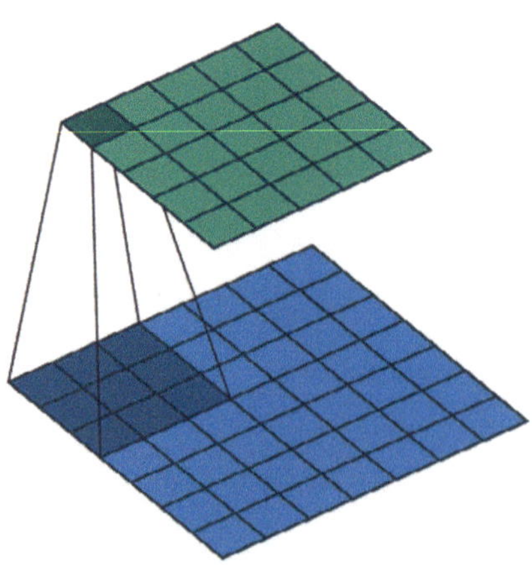

Fig. 4 3×3 filters.

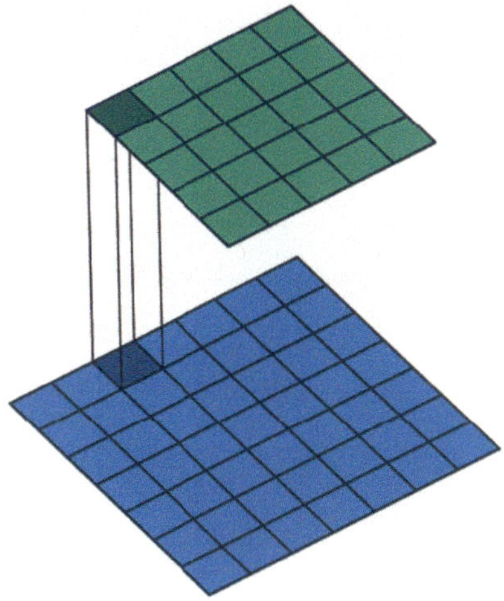

Fig. 5 1×1 filters.

Strategy 2

Strategy 2 is used to lessen the total inputs in abide of 3×3 filters.

- Fewer inputs to Conv layers result in fewer parameters.
- It is achieved by using only 1×1 filters before the 3×3 Conv layers.
- It is called the squeeze layer (description in next section).
- Total parameters in a 3×3 Conv layer = (total input channels) (total filters) (3×3).

Strategy 3

Strategy 3 is used in the down-sample late, which are in the grid because of that the convolution layers will retain more number of the awaking maps.

- It mostly makes a smaller number of parameters and maximizes accuracy.
- Delaying down sampling late in the network creates more extensive activation of feature maps.
- Leaving from more traditional architectures like the VGG network that use early down sampling.
- Large activation maps result in a higher classification accuracy given the same number of parameters.

(b) Fire Module

In the fire module, there is various squeeze convolution layer, and it consists of only 1×1 filters, which provide it into an expand layer

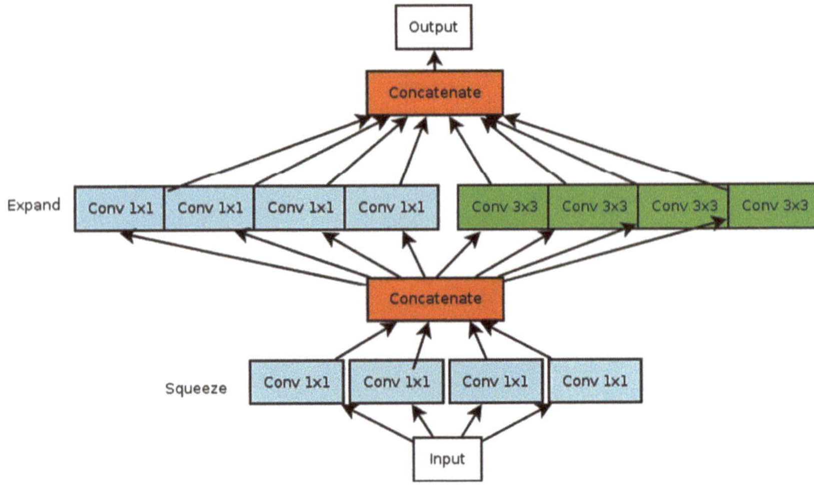

Fig. 6 Fire module.

consisting of a mixture of 1 × 1 and 3 × 3 convolution filters, which is shown in Fig. 6. Here, three different tunable dimensions (hyper-parameters) are in a Fire module: s1 × 1, e1 × 1, and e3 × 3. The hyper-parameters (i.e., s1 × 1) in a fire module has many filters which are utilized in the layer, which is called the squeeze layer (all 1 × 1) has a total of 1 × 1 in expanding layer then here in this stage, e3 × 3 is the total of 3 × 3 filters which is also used in expanding layer. Here, the Fire module is used and deposited in the s1x1which must be smaller than (e1x1 + e3 × 3). Therefore, the squeeze layer supports restricting the total input channels of the 3 × 3 filters by using Strategy 2.

(c) SqueezeNet Architecture

The SqueezeNet Convolutional Neural Network (CNN) architecture, illustrated in Fig. 7 in which the SqueezeNet begins beside a standalone convolutional layer (i.e., conv1), which will come after with a totally of 8 Fire modules (i.e., fire2–9) and it will end with a final convolutional layer (i.e., conv10). Here, total filters slowly rise in each fire module, which starts close to finishing the network. The max pool with a stride of 2 below these layers in SqueezeNet: conv1, fire4, fire8, and conv10; these layers are almost late placements of the pooling are in each of the Strategy 3.

Layers breakdown has eleven layers:

- The first layer is the regular Convolution layer.
- The second layer up to the ninth layer is the Fire module layer (squeeze + expand layer).

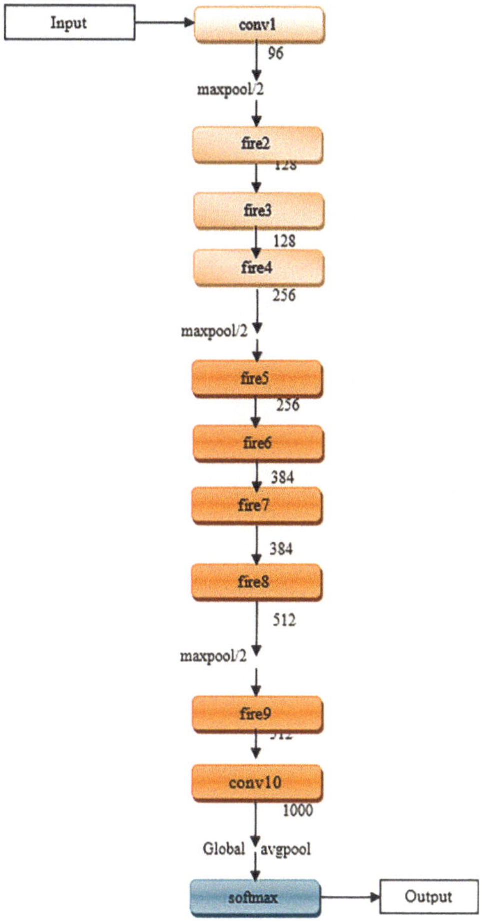

Fig. 7 SqueezeNet architecture.

- The tenth layer is also the regular convolution layer.
- The eleventh layer is the Softmax layer.

3.3 Feature extraction

Feature extraction is the procedure of planning a current Convolutional Neural Network design which is a very monotonous step, and it needs the large capacity of the information and the assets during the implementation. The pre-coached designs like AlexNet will assimilate during the transfer learning. Here, AlexNet works are dependent on the architecture of CNN for transfer learning. Initially, the pre-coached AlexNet designs along the ImageNet dataset force are extracted for the feature extraction along the

ImageNet dataset with the comprehend filters of the Convolutional Neural Network architectures; by using these steps, feature coordinate is collected. The pre-coached CNN pattern has the initial category of layers consisting of the least number of features with crucial complementary details similar to the edges. Finally, completely connected layers of the pre-coached network are extracted. The utterly connected layer neuron is now added to the head of the convolutional layers. They are used for gathering the removed features from the convolutional layers. Finally, the inspection of the various completely connected layers has an adequate perception as the decision.

3.4 Residual network (ResNet) classification

ResNet has been designed to create networks that are ultra-deep. ResNet falls under the deep learning category. ResNet is used not only for deep learning but also for more applications like natural language processing. ResNet was designed with several layers of nodes, such as making 34th layer, 50th layer, 101st layer, 152nd layer, and even 1202nd layer of nodes with different layers. ResNet50, the famous convolutional network, is the network with 49 convolutional layers and one fully-connected layer at the network's bottom. To get the total weight and MAC to the entire network, we need 25.5 million pounds and 3.9 million pounds. The ResNet architecture is shown as the basic block diagram, as seen in the colorful illustration in Fig. 8.

Residual Network (ResNet) is a conventional feed-forward lattice that has the remaining connection. The residual layer's output is explained by the bases in the outputs by the precursory layer, explained as $xl-1$. $F(xl-1)$ is the output behind executing different performances. For example, the convolution with dissimilar sizes of the filters, Batch Normalization (B.N.) is pursued by the stimulation responsibility, like a ReLU $xl-1$. The accomplishment output of the residual of the unit is xl that the following equation might express as,

$$xl = F(xl - 1) + xl - 1 \quad (1)$$

Residual network (ResNet) contains lots of the fundamental residual blocks. The functions present in the residual block could be differed based on the non-identical structural design of the residual networks. The residual network's comprehensive description was suggested, and then the surrogate advanced residual network perspective is called as the accumulated residual metamorphose was produced.

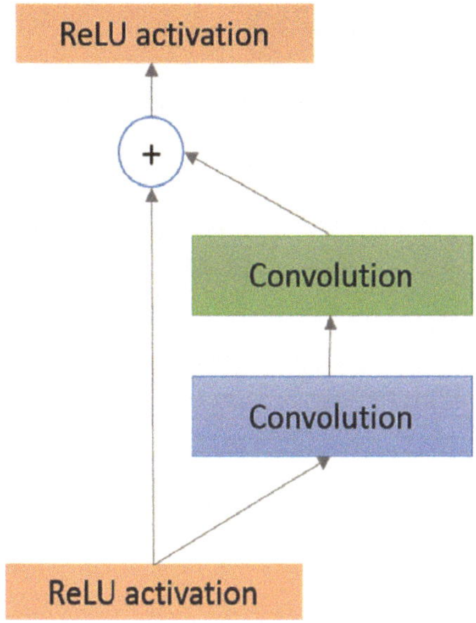

Fig. 8 Block diagram of the ResNet architecture.

Newly, some interchangeable alternative of the residual designs is initiated, and it depends on the Residual Network architecture. Furthermore, various developed architectures are integrated by the establishment of the Residual units. The fundamental imaginary illustration of the Inception Residual unit is exhibited in the following Fig. 9.

4 Simulation results and discussion

As the validation of the Proposed work, the input is given, and the parameters measured play a significant role. Some sample data set has been used as input. The simulation has been resulted using MATLAB based on the parameter like high precision, Dice scores, specificity, and precision. The following Fig. 10 represents the sample C.T. lungs image tested with this proposed work.

The RGB image is converted to a Gray image. The input images consume more memory to process. Therefore the image is converted into grayscale where only one band will represent the pixel value that varied from 0 to 255. The following Fig. 11 shows the grayscale image.

The preprocessed image is the extraction of the portion of the lung from the X-ray scan input along with the other pixels are given as "0." In the

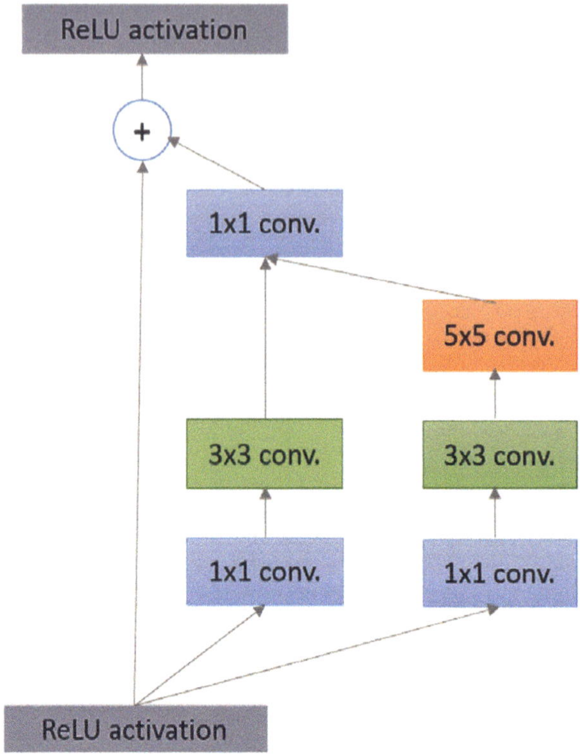

Fig. 9 Block diagram for inception residual unit.

preprocessed image, the image contrast is improved, and the noises are removed. After grayscale conversion, improve the contrast of the image by using an average filter. The following Fig. 12 represents the pre-processing image.

After pre-processing, the image is contoured. The process of the initial contour is used to contour the lungs separately in the C.T. image. The following Fig. 13 represents the Initial contour.

After contouring, the image has been iterated. The iteration stage is the operation of manipulating individual components in the digital structure. In order to operate all of the pixels, we need to be able to visit all the rows and the columns entirely in the image. The following Fig. 14 represents the iteration image.

The final boundary mask image will convert the segmentation gray image to the black and white image. This process separates the lung image from the C.T. image. The following Fig. 15 represents the final boundary mask.

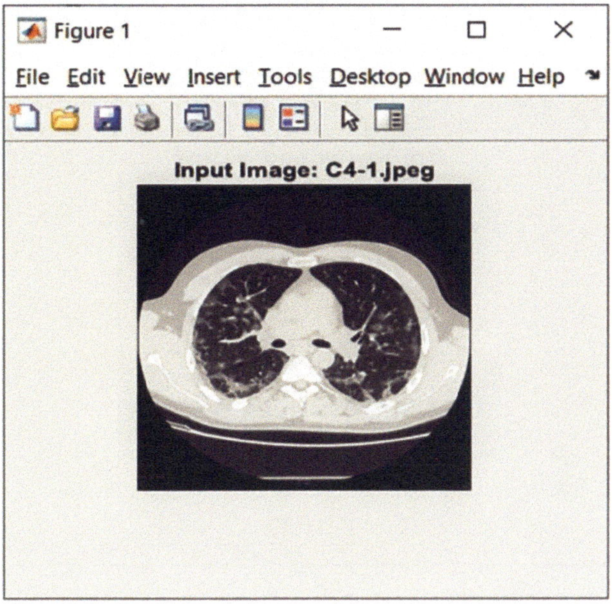

Fig. 10 Input image.

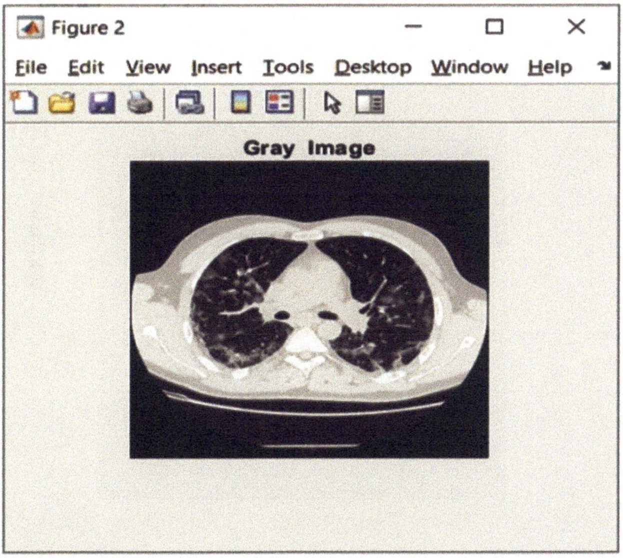

Fig. 11 Grayscale image.

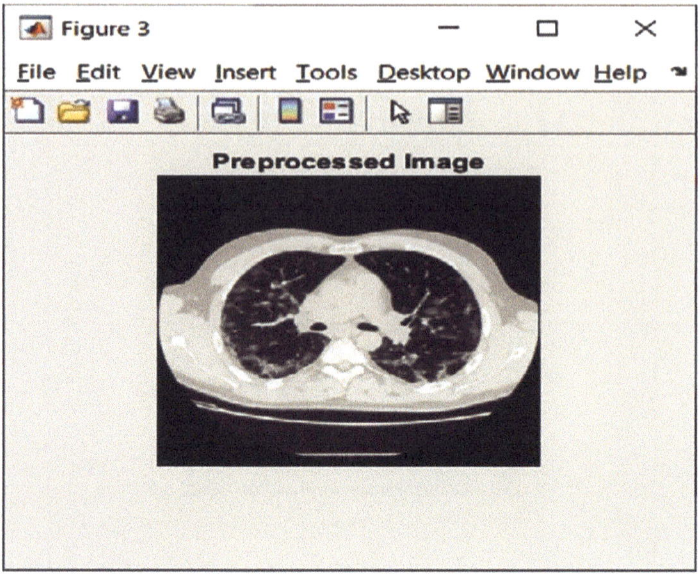

Fig. 12 Pre-processed image.

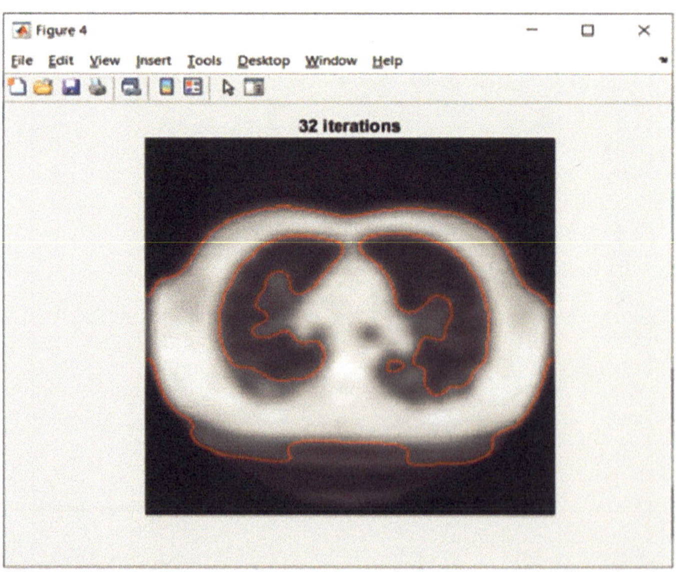

Fig. 13 Image at the initial contour.

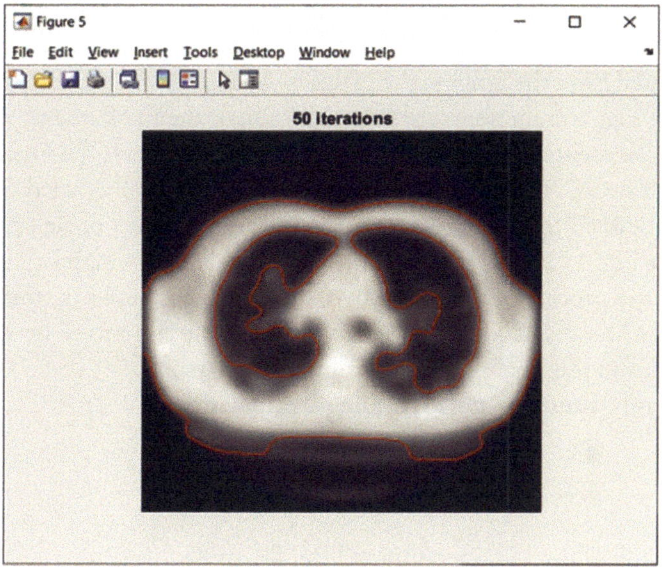

Fig. 14 Iteration image.

Fig. 15 Final boundary mask image.

After the final boundary mask image process, the Segmentation takes place by SqueezeNet CNN. In this segmentation process, the infected region is segmented from the lungs in the image by SqueezeNet Segmentation. The following Fig. 16 represents the Segmentation of the mask image.

After Segmentation, the layer processing step compares the image to the already trained C.T. images and produces the result. CNN-based classification is processed to classify whether the image is positive or negative. The following Fig. 17 shows the first convolutional layer's weights.

After the process of the image by Convolutional Network, the result is displayed. The result shows whether COVID-19 is positive or negative. Fig. 18 shows the final predicted output of the system.

Formula used for performance evaluation:

$$\text{Sensitivity} = \text{TP}/\text{P} \tag{2}$$

$$\text{Specificity} = \text{TN}/\text{N} \tag{3}$$

$$\text{Precision} = \text{TP}/(\text{TP} + \text{FP}) \tag{4}$$

$$\text{Accuracy} = (\text{TP} + \text{TN})/(\text{P} + \text{N}) \tag{5}$$

Dice score shall be calculated from the below formula,

$$\text{Dice score} = 2 \times \text{number of true positives}/(2 \times \text{number of true positives} + \text{number of false positives}) + \text{number of false negatives} \tag{6}$$

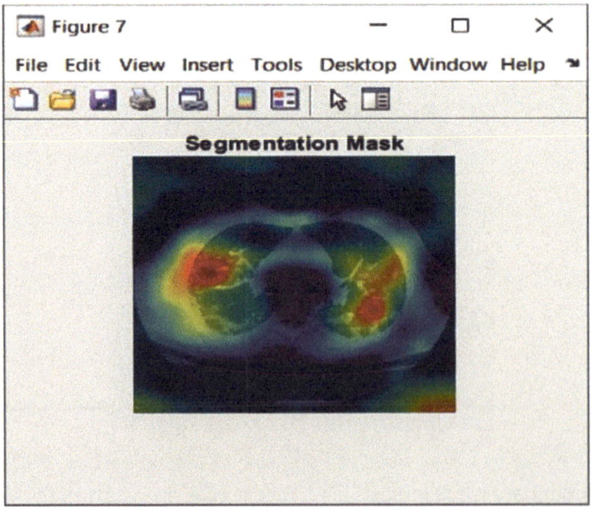

Fig. 16 Segmentation mask image.

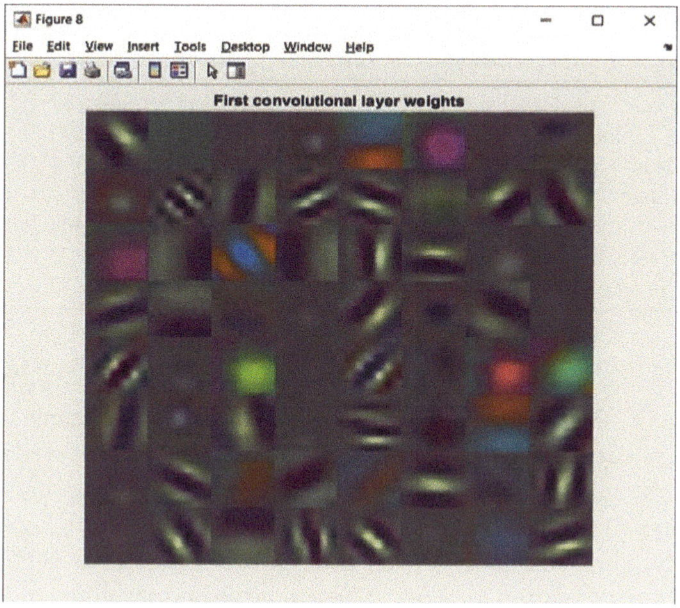

Fig. 17 CNN processing image large weights.

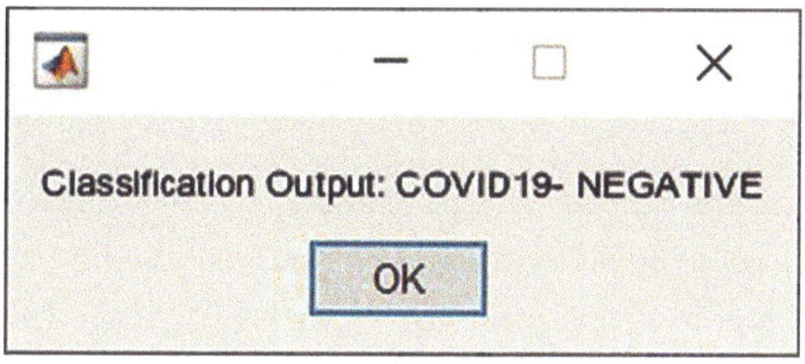

Fig. 18 Classification output.

Where,
P—condition positive
N—condition negative
TP—True positive
TN—True negative
FP—false positive
FN—false negative

Here, P denotes the number of real positive cases in the data. N represents the number of real negative cases in the data. TP is equivalent with hit. TN denotes equivalent with correct rejection. FP represents equivalent with underestimation and FN denotes equivalent with overestimation.

Table 1 Performance evaluation.

Model	Dice	Sen.	Spec.	Prec.	Accu.
U-Net	0.309	0.676	0.837	0.267	91.7
Attention-UNet	0.468	0.725	0.934	0.393	92.9
Gated-UNet	0.445	0.673	0.958	0.376	93.3
Dense-UNet	0.412	0.609	0.979	0.417	94.8
U-Net++	0.447	0.876	0.927	0.367	92.1
Squeeze-Net	0.578	0.871	0.977	0.503	97.8
Res-Net(proposed)	0.599	0.867	0.979	0.517	98.2

Table 1 shows the performance comparison of different existing methods with the proposed method. The specificity of the proposed method is 0.977 out of 1. The accuracy of the result produced from Fig. 17 is 98.2%. The proposed method accuracy has been ensured based on high precision, specificity, and Dice scores, even though the sensitivity is slightly low compared with existing methods. The specificity and precision are directly proportional to accuracy. Table 1 performance evaluation parameters rounding the scale value is 1, except accuracy mentioned in the percentage.

5 Conclusion

Thus an automatic detection of coronavirus 2019 infection in the lung by using segmentation process from C.T. images has been simulated. The named deep residual network in Convolutional Neural Network (CNN) is used and the definite reverse attention and the distinct edge attention for developing the recognition of the affected area in the lung are applied, with the help of chest C.T. slices. This structure holds admirable possible which can be sought during the evaluation in detecting COVID-19, for example, computing the affected area, observing the longitudinal infection difference, and aggregation disguising transforming. Datasets were performed as input data to the SqueezeNet network using the image processing techniques. The network achieved is of higher accuracy. Thus the proposed model can recognize the recipients with less power and compare the lung and the affected area of the normal tissues. The Simulation results revealed that the Accuracy had attained it peak compared to the existing models.

References

Abdel-Basset, M., Chang, V., & Mohamed, R. (2020). HSMA_WOA: A hybrid novel slime mould algorithm with whale optimization algorithm for tackling the image segmentation problem of chest X-ray images. *Applied Soft Computing*, *95*, 106642.

Abdel-Basset, M., Chang, V., & Nabeeh, N. A. (2021). An intelligent framework using disruptive technologies for COVID-19 analysis. *Technological Forecasting and Social Change*, *163*, 120431.

Abitha, N., Babysha, P., Devisri, S., & Kumar, P. (2020). Application of DL/ML in diagnosis in medical imaging. *Journal on Intelligent Systems & Robotics Insights & Transformations*, *4*(1) (ISSN: 2581-5636 (online)).

Ai, T., Yang, Z., Hou, H., Zhan, C., Chen, C., Lv, W., et al. (2020). Correlation of chest C.T. and RT-PCR testing for coronavirus disease 2019 (COVID-19) in China: A report of 1014 cases. *Radiology*, *296*(2), E32–E40.

Balaji, S., Balamurugan, B., Kumar, T. A., Rajmohan, R., & Kumar, P. P. (2021). A brief survey on AI-based face mask detection systems for public places. *Irish Interdisciplinary Journal of Science & Research (IIJSR)*, *5*(1), 108–117.

Bertram, K. M., Truong, N. R., Smith, J. B., Kim, M., Sandgren, K. J., Feng, K. L., et al. (2021). Herpes simplex virus type 1 infects Langerhans cells and the novel epidermal dendritic cell, Epi-cDC2s, via different entry pathways. *PLoS Pathogens*, *17*(4), e1009536.

Chowdhury, M. E., Rahman, T., Khandakar, A., Mazhar, R., Kadir, M. A., Mahbub, Z. B., et al. (2020). Can AI help in screening viral and COVID-19 pneumonia? *IEEE Access*, *8*, 132665–132676.

Cohen, J. P., Morrison, P., Dao, L., Roth, K., Duong, T. Q., & Ghassemi, M. (2020). *Covid-19 image data collection: Prospective predictions are the future.* arXiv preprint arXiv:2006.11988.

Dhiman, G., Chang, V., Kant Singh, K., & Shankar, A. (2021). Adopt: Automatic deep learning and optimization-based approach for detection of novel coronavirus covid-19 disease using x-ray images. *Journal of Biomolecular Structure and Dynamics*, 1–13. https://doi.org/10.1080/07391102.2021.1875049.

Fan, D. P., Zhou, T., Ji, G. P., Zhou, Y., Chen, G., Fu, H., et al. (2020). Inf-net: Automatic covid-19 lung infection segmentation from ct images. *IEEE Transactions on Medical Imaging*, *39*(8), 2626–2637.

Fang, Y., Zhang, H., Xie, J., Lin, M., Ying, L., Pang, P., et al. (2020). Sensitivity of chest C.T. for COVID-19: Comparison to RT-PCR. *Radiology*, *296*(2), E115–E117.

Gokulan, S., Narmadha, S., Pavithra, M., Rajmohan, R., & Ananthkumar, T. (2020, July). Determination of various deep learning parameter for sleep disorder. In *2020 international conference on system, computation, automation and networking (ICSCAN)* (pp. 1–6). IEEE.

Hernigou, J., Cornil, F., Poignard, A., El Bouchaibi, S., Mani, J., Naouri, J. F., et al. (2020). Thoracic computerised tomography scans in one hundred eighteen orthopaedic patients during the COVID-19 pandemic: Identification of chest lesions; added values; help in managing patients; burden on the computerised tomography scan department. *International Orthopaedics*, *44*(8), 1571–1580.

Huang, C., Wang, Y., Li, X., Ren, L., Zhao, J., Hu, Y., et al. (2020). Clinical features of patients infected with 2019 novel coronavirus in Wuhan, China. *The Lancet*, *395*(10223), 497–506.

Jeyaranjani, J., Rajkumar, T. D., & Kumar, T. A. (2021). Coronary heart disease diagnosis using the efficient ANN model. *Materials Today: Proceedings*, *1*, 1–5. https://doi.org/10.1016/j.matpr.2021.01.257.

Jin, D., Xu, Z., Tang, Y., Harrison, A. P., & Mollura, D. J. (2018, September). CT-realistic lung nodule simulation from 3D conditional generative adversarial networks for robust

lung segmentation. In *International conference on medical image computing and computer-assisted intervention* (pp. 732–740). Cham: Springer.
Keshani, M., Azimifar, Z., Tajeripour, F., & Boostani, R. (2013). Lung nodule segmentation and recognition using SVM classifier and active contour modeling: A complete intelligent system. *Computers in Biology and Medicine, 43*(4), 287–300.
Kumar, T. A., John, A., & Kumar, C. R. (2020). *2 IoT technology and applications. Internet of Things* (p. 43). Germany: De Gruyt.
Kumar, T. A., Selvi, S. A., Rajesh, R. S., Perumal, P. S., & Stalin, J. (2021). Limos—Live patient monitoring system. In *Handbook of artificial intelligence in biomedical engineering* (pp. 311–330). Apple Academic Press.
Pavithra, M., Rajmohan, R., Kumar, T. A., & Ramya, R. (2021). Prediction and classification of breast cancer using discriminative learning models and techniques. In *Machine vision inspection systems, volume 2: Machine learning-based approaches* (pp. 241–262). India: Scrivener Publishing LLC.
Rajinikanth, V., Dey, N., Raj, A. N. J., Hassanien, A. E., Santosh, K. C., & Raja, N. (2020). *Harmony-search and otsu based system for coronavirus disease (COVID-19) detection using lung C.T. scan images.* arXiv preprint arXiv:2004.03431.
Rajmohan, R., Kumar, T. A., Pavithra, M., & Sandhya, S. G. (2020). 11 Blockchain. In *177. Blockchain technology: Fundamentals, applications, and case studies.* United States: CRC Press.
Samuel, T. A., Pavithra, M., & Mohan, R. R. (2021). LIFI-based radiation-free monitoring and transmission device for hospitals/public places. In *Multimedia and sensory input for augmented, mixed, and virtual reality* (pp. 195–205). IGI Global.
Shi, F., Wang, J., Shi, J., Wu, Z., Wang, Q., Tang, Z., et al. (2020). Review of artificial intelligence techniques in imaging data acquisition, segmentation and diagnosis for covid-19. *IEEE Reviews in Biomedical Engineering, 14*, 4–15. 2021 https://doi.org/10.1109/RBME.2020.2987975.
Solomon, M. D., McNulty, E. J., Rana, J. S., Leong, T. K., Lee, C., Sung, S. H., et al. (2020). The Covid-19 pandemic and the incidence of acute myocardial infarction. *New England Journal of Medicine, 383*(7), 691–693.
Srimathi, B., & Ananthkumar, T. (2020). Li-Fi based automated patient healthcare monitoring system. *Indian Journal of Public Health Research & Development, 11*(2), 393–398.
Suresh Kumar, K., Radha Mani, A. S., Sundaresan, S., & Ananth Kumar, T. (2021). Modeling of VANET for future generation transportation system through edge/fog/cloud computing powered by 6G. In *Cloud and IoT-based vehicular ad hoc networks* (pp. 105–124). United States: : John Wiley & Sons, Inc.
Wang, C., Horby, P. W., Hayden, F. G., & Gao, G. F. (2020). A novel coronavirus outbreak of global health concern. *The Lancet, 395*(10223), 470–473.
Wang, S., Zhou, M., Liu, Z., Liu, Z., Gu, D., Zang, Y., et al. (2017). Central focused convolutional neural networks: Developing a data-driven model for lung nodule segmentation. *Medical Image Analysis, 40*, 172–183.
Ye, Z., Zhang, Y., Wang, Y., Huang, Z., & Song, B. (2020). Chest CT manifestations of new coronavirus disease 2019 (COVID-19): A pictorial review. *European Radiology, 30*(8), 4381–4389.
Zhao, S., Huang, Q., Tang, Y., Yao, X., Yang, J., Ding, G., et al. (2021). *Computational emotion analysis from images: Recent advances and future directions.* arXiv preprint arXiv:2103.10798.
Zhou, Z., Siddiquee, M. M. R., Tajbakhsh, N., & Liang, J. (2018). Unet++: A nested u-net architecture for medical image segmentation. In *Deep learning in medical image analysis and multimodal learning for clinical decision support* (pp. 3–11). Cham: Springer.

Index

Note: Page numbers followed by *f* indicate figures and *t* indicate tables.

FSC
www.fsc.org
MIX
Papier | Fördert
gute Waldnutzung
FSC® C083411

Kolibri 360 GmbH
Ferdinand-Jühlke-Straße 7
99095 Erfurt, Deutschland
produktsicherheit@kolibri360.de